THE PREHISTORIC NATIVE AMERICAN ART OF MUD GLYPH CAVE

EDITED BY

Charles H. Faulkner

THE UNIVERSITY OF TENNESSEE PRESS

Library of Congress Cataloging-in-Publication Data

The Prehistoric native American art of Mud Glyph Cave.

  Bibliography: p.
  Includes index.
  Contents: Mud Glyph Cave / by Howard H. Earnest, Jr.
and Charles H. Faulkner—The underground setting / by Art
Cathers—Recording the glyphs / by Bill Deane—[etc.]
  1. Mud Glyph Cave (Tenn.) 2. Indians of North
America—Tennessee—Antiquities. 3. Cave-drawings—
Tennessee. 4. Petroglyphs—Southern States. 5. Tennessee—
Antiquities. I. Faulkner, Charles H.
E78.T3P74   1986   976.8'01   86–1697
ISBN 0-87049-505-4 (alk. paper)

CONTENTS

ILLUSTRATIONS

PLATES

FIGURES

TABLES

ACKNOWLEDGMENTS

The success of the Mud Glyph Cave Project is due to the interest, support, and labor of many people. I would especially like to thank the National Geographic Society for the grant (No. 2421-82) that made the research in Mud Glyph Cave possible, and the landowners, who unfortunately must remain anonymous for security reasons, for allowing us access to their property and for protecting the cave. I am also particularly grateful to Walter Merrill, who first discovered the glyph gallery, and Howard Earnest, who brought it to my attention. The cave was surveyed, mapped, and photographed with the assistance of Art Cathers, Howard Earnest, Chuck Meier, Jim Nicholls, Melvin Whaley, Gary Wilkins, and John Yust. Bill Deane deserves special recognition for coordinating this work underground. Gary Crites identified the botanical remains, and Robert Stuckenrath, Radiation Biology Laboratory, Smithsonian Institution, provided the radiocarbon dates. Professional consultants on the project were B. Bart Henson, Jon Muller, Louise Robbins, Patty Jo Watson, and Ronald Wilson. The photographs in Chapters II and VI were made by Bill Deane; those in Chapter VII were made by B. Bart Henson. The cave maps were originally drawn by Jim Nicholls; these maps (figures 1, 2) were modified by Terry Faulkner. Pam Poe typed the final draft of the manuscript. Just as the field work was an enjoyable joint effort, so was the writing of this book. It would not have been possible without the dedication of the authors who share my fascination with Mud Glyph Cave.

Charles H. Faulkner

Department of Anthropology
University of Tennessee, Knoxville

THE PREHISTORIC
NATIVE AMERICAN ART OF
MUD GLYPH CAVE

HOWARD H. EARNEST, JR.
AND CHARLES H. FAULKNER

I

Mud Glyph Cave:
Its Discovery and
Archaeological Study

On a cold day in early 1980, Walter Merrill and Lloyd Hamby crawled into a small cave opening in an East Tennessee hillside. They made their way into the narrow cave, not deterred by a tight, muddy crawlway, tiny keyhole passages in the rock, and knee-deep water. Some 120 meters from the entrance they reached a higher passage, well above the small stream they had been following. Their flashlights revealed incised lines covering the thick mud coating of the walls of this higher passage. Merrill and Hamby had explored many caves in the area. They knew that these limestone solution caves often had mud-coated walls, and that many such walls were defaced with modern graffiti. This cave, they soon realized, was different. They saw none of the usual initials, dates, and profanity, only some recognizable human and animal figures, and abstract lines.

Merrill and Hamby agreed to keep their discovery quiet and returned to the cave a few weeks later to take a better look at the mud drawings and to confirm the absence of modern graffiti. Both men worked for the United States Forest Service on the local Ranger Districts of the Cherokee National Forest. Merrill had frequently worked with the Forest Service Archaeologist Howard Earnest, Jr. The next time they worked together, in March 1980, Merrill told Earnest about the cave and offered to show him the drawings at the first opportunity. Earnest's usual professional skepticism about such reports was tempered by his trust in Merrill's observations. Merrill had a keen interest in local history and geology, and this had made him an important informant about the National Forest sites which Earnest was responsible for.

The promised trip to the cave took some time to schedule because the cave was on private property, and the visit could not be made on a workday. Finally on a Sunday afternoon in October 1980 another expedition was arranged by Merrill and Hamby. In addition to Earnest, they invited Herbert Bell, another Forest Service worker who was an accomplished photographer, to record the mud drawings on film. Forest Service District Ranger Guy Thurmond, Bell's son Anthony, and a friend of Earnest, Martin Aiken, made up the rest of the team. Though not sharing Merrill's and Hamby's enthusiasm for crawling through small holes in the ground, the group reached the passage they sought. The first impression that Earnest had upon viewing the cave drawings was that the mud coating and the drawings *looked* very old. The surface of the clay, including the incised lines, was darkened with soot or mineral deposits. Places where the mud had fallen from steep walls showed its natural color to be a much lighter orange-brown. All doubt as to the antiquity of the work was removed from Earnest's mind when he saw a high steep ledge, out of reach from the floor, covered with complex stylized figures reminiscent of late prehistoric shell gorget engraving in the region. As they left the cave, the group talked about how this amazing body of prehistoric art came to be. Discussing the local archaeological record and more recent history with Merrill, Earnest concluded that the most incredible thing was the preservation of the drawings, not the fact that they had existed in the first place. It was *not* surprising that prehistoric Indians had used this small cave. It lies at one end of a fertile valley, covered with rich alluvial soils, today a patchwork of prosperous farms. Surrounded by poor shaly ridges and rugged mountains in the Ridge and Valley physiographic province of East Tennessee, this valley undoubtedly attracted Archaic Period hunter-gatherers as well as later agriculturalists. The area has not been surveyed for archaeological sites, but two small mounds, probably Late Woodland Period burial mounds, are found about a mile from the cave. Local residents say that another mound was razed a few years ago. They also report finding artifacts, including pottery, in plowed fields in the valley. The pottery and mounds indicate permanent populations during the Woodland Period and possibly later.

In the immediate vicinity of the cave entrance, there is no evidence of permanent Indian occupation. Here the valley is narrowing and the large creek which drains the area begins to wind for several miles through a relatively narrow gorge. On a small strip of alluvium between the creek and base of the ridge where the entrance lies, a few scattered chert and quartz flakes have been found, evidence of tool maintenance by passing hunters.

Not until the early part of this century did the area outside the cave attract permanent residents. White farmers built a frame house at the base of the hill near the stream flowing from the cave. They built a springhouse and channeled the stream with stone retaining walls. The dry upper cave entrance was used as a root cellar. Despite its importance to them, there is no evidence that these residents explored the cave beyond this entrance-room storage area. The tight crawlway just beyond this room could have been plugged with alluvial clay in the past, or it may have been rocked up to keep raccoons and the like from the stored foodstuffs.

Wanting to spend more time inspecting the cave and drawings, Earnest planned another trip to the cave. He and Michael Barber, archaeologist for the Forest Service in Virginia, spent several hours in the cave on their Veterans' Day holiday. Earnest and Barber took measurements and took more photographs. They also searched the passage for artifacts but found only charred rivercane and wood. Barber concurred with Earnest's opinion that the drawings were not modern, that they were extremely important and extremely fragile. Systematic investigation and recording of the drawings was urgently needed.

Unable to apply for the necessary funds to do the work from his government post, Earnest contacted his undergraduate mentor Professor Charles Faulkner of the University of Tennessee at Knoxville. Armed with Herbert Bell's excellent black-and-white photographs and slides taken by Barber and himself, Earnest had no trouble convincing Faulkner that a discovery of great significance had been made. They also agreed it was imperative that a grantor be found as soon as possible to fund the study of these drawings before they were destroyed by curiosity seekers and vandals.

In January 1982 a grant was received from the National Geographic Society to study what is now called "Mud Glyph Cave." Fieldwork began on February 23, 1982, and continued through October of that year with

Charles Faulkner as principal investigator. The research design included four phases of site preservation and data recording. With the full cooperation of the landowners, the cave entrance was gated to protect the fragile glyphs. Fortunately, access to the cave was restricted before serious vandalism could occur.

The next phase of research included the complete mapping of the cave and an archaeological survey of the cave floor for cultural remains of the people responsible for this cave art. The entire cave was mapped at a scale of 35 mm = 7 meters by Bill Deane, Art Cathers, Walter Merrill, and Jim Nicholls, with the final map being drafted by Nicholls. A detailed map of the glyph passage was also produced by Nicholls at a scale of 35 mm = 5 meters, featuring the location of the major glyphs on the walls and cultural features on the floor. The glyph passage was staked at one meter intervals with small numbered aluminum pins to facilitate mapping and to be used as reference points for photography and the floor survey.

The archaeological survey was conducted under the direction of Howard Earnest with Gary Wilkins assisting. The cave floor was searched for discarded artifacts, features such as hearths were located and mapped, soil samples were taken from basin-shaped depressions in the cave floor, and charcoal samples for C–14 dating and plant identification were collected from concentrations of torch charcoal and from the contents of hearths.

After the cave was mapped and surveyed, a complete photographic record was made of both walls of the glyph passage by professional cave photographer Bill Deane. Color slides and black and white prints of the glyph-covered walls were provenienced by the one meter grid stakes on the cave floor. In addition, color slides were made of the cave entrance, of other passages in the cave, of researchers gathering data, and of individual glyphs.

The next phase of research in Mud Glyph Cave involved visits by archaeological consultants who were considered experts in cave archaeology and the art of primitive people. On May 17 and May 19, 1982, the cave was visited by Jon Muller of Southern Illinois University, Bart Henson of Huntsville, Alabama, and a University of Tennessee at Knoxville research team for the purpose of comparing the glyphs to motifs on prehistoric artifacts and rock shelter and cave walls in the southeastern United States. A team of consultants again entered the cave on July 23,

1982, to establish the nature of cultural activities in the cave other than the artwork, to determine if animals other than man had marked the cave walls, and to make casts of some of the tool impressions in the mud. Consultants on this trip included Patty Jo Watson, Washington University; Ronald Wilson, University of Louisville; Louise Robbins, University of North Carolina at Greensboro; and Bart Henson. The UTK research team of Charles Faulkner, Bill Deane, Jim Nicholls, Art Cathers, and Chuck Meier served as guides.

The last research phase was the verification of design elements, motifs and structure of the glyphs identified on enlarged black and white photographs of the glyph gallery walls. On June 23, 1984, John Muller and Jeannette Stephens, assisted by Art Cathers, Charles Faulkner, and Chuck Meier, spent eleven hours in the glyph passage comparing the elements and motifs identified on copies of the photographs and those actually present on the walls to make sure no obscure lines were missed. It was found that the photographs accurately revealed all elements on the walls and had been the most effective, as well as efficient means, to record the glyphs.

II

The Underground Setting: Survey and Mapping of Mud Glyph Cave

The initial phase of the archaeological research at Mud Glyph Cave began with the production of a detailed map of the cave. This was intended to serve several purposes.

On any archaeological site, permanent datum points are required for reference, and such datum points were established during the survey of the cave. The map itself, which showed these datum points, served as a reference for those working at the cave and would aid in describing the site to other interested researchers. In view of the overall fragility of the site, and the virtual inevitability of damage to the glyphs with continuing visits to the cave, the latter function would be crucial. Much of the early work at Mud Glyph Cave, including the mapping, was aimed at describing the cave in sufficient detail so that future research could be conducted off-site for the most part.

Cave mapping usually involves more art than science. Cave passages are seldom simple tunnels with straight, vertical walls and flat ceiling and floor that are easily portrayed on a two-dimensional map. More often the vagaries of the processes that form and modify solution caves combine to produce passages that greatly vary in shape and size and meander about and pass over, under, around and through other passages. The resulting three-dimensional space can be extremely difficult to portray on paper, and many measurements are required to portray it accurately. In practice, time limitations permit only a few measurements, and those can only define the overall layout of the cave and the size of the passages at a few points. The majority of the information on a typical cave map must be derived from sketches made in the cave.

Further compromises are necessary in the few measurements that can be made. Cave surveyors work in total darkness, must wade or swim through water, climb steep rock faces, crawl through low passages, and squeeze through tight openings to make their measurements; their clothing usually becomes caked with mud. Techniques that allow extreme precision in surveys above ground can be very difficult to use underground, and cave surveys generally sacrifice some accuracy in the interest of speed and flexibility.

Mud Glyph Cave, fortunately, is fairly simple in layout and not too difficult to traverse. The small entrance, which has now been extensively modified and gated, opens into a low room, used at one time to store sweet potatoes. A stooping passage (Plate I), interrupted by a belly crawl, leads to the first keyhole, an awkwardly situated opening in the right wall of the passage. Beyond the keyhole one encounters the cave stream, which disappears into a small hole downstream, reappearing at a spring below the entrance. Heading upstream, a narrow slanting passage (Plate II) leads to the second keyhole, which is a low opening in the left wall, just above the stream, and is also awkward to negotiate. Beyond is a large walking stream passage, which soon leads to a fork (Plate III). Directly ahead, a steep mud slope leads up to the glyph gallery (Plate IV), the large walking passage that contains most of the evidence of prehistoric visitation.

The glyph gallery is about 120 meters from the cave entrance. The 96 meter long gallery passage was apparently formed as the limestone was dissolved and eroded by flowing water—the same process that is still occurring in the lower passage today. As the water continued to fill this passage after it reached its present configuration, probably at a later period of time, a fine silty clay was deposited on the walls, reaching a thickness of 10 cm on some areas of the walls. When the water finally receded, the stream was confined to the lower passage; it has not flooded the glyph gallery since the glyphs were drawn hundreds of years ago. The lower stream passage still floods during periods of heavy rainfall, and since our research in the cave this water has lapped over the floor at the entrance to the glyph passage. The high humidity and constant temperature in the cave have kept this veneer of mud soft and workable

PLATE I. Stooping passage and belly crawl just inside entrance.

and have provided a perfect medium for artistic expression by the aboriginal visitors (Plate V).

To the left the stream passage continues, paralleling the glyph gallery. Two windows in the gallery overlook the stream, and at its far end the gallery rejoins the stream passage, via another steep slope. The stream passage continues past some petroglyphs on the left wall and then abruptly veers to the right and becomes much lower. Just beyond the turn, a climb of about 7 meters leads into a large upper room. Below the climb, the stream passage continues as a low wet crawl, ending at a siphon. During the survey, cavers swam through the siphon, finding another section of the large stream passage and reaching the apparent end of the cave.

A map of the cave is shown in Figures 1 and 2. Given the nature of the cave, a simple plan view, showing the passages from above and using cross-sections to augment the basic map seemed to be the best way to illustrate the cave. Before the map could be prepared, however, the survey had to be initiated.

Cave surveys are relatively simple in conception. Markers, or survey

PLATE II. Narrow slanting passage above stream.

PLATE III. Walking stream passage.

PLATE IV. Steep mud slope leading to glyph gallery.

PLATE V. Fireplace room and high ledge.

points, are placed sequentially at suitable locations throughout the cave, each marker within sight of one preceding it and generally a few meters distant, depending on the size and character of the intervening cave passage. As each new point is chosen, measurements are made to define the orientation and length of the imaginary vector, or survey shot, linking it to the preceding point. A sketch is drawn, showing the details of the section of cave passage near each point, in plan view and in cross-section, and ancillary measurements are made to specify the height and width of the passage and the location of the point within the passage.

Surveys are usually started at a cave entrance and proceed into the cave via the largest or easiest available passage, later branching out to cover side passages. Eventually, the survey is extended to cover the entire cave, often after several trips.

Before the map can be drawn, data from the survey must be reduced to useful form. Coordinate axes are chosen, generally using magnetic north and east as horizontal axes, with the remaining axis being vertical. The measurements from each survey shot are then converted trigonometrically to give displacements along these three axes. One of the survey points is picked to be a base point, corresponding to the origin of the coordinate system, and a convenient scale for the map is chosen. Since each point in the survey is linked by a series of survey shots to the base, the position of each point can be calculated by summing the displacements corresponding to the appropriate shots. The result, plotted on graph paper with lines sketched in to show the survey shots linking the points, is a line plot of the cave—a skeletal form of the cave map. Using the sketches and notes recorded at the cave, the walls and details of the cave passage are then filled in. Finally, with the addition of cross-sections and various interpretive aids, the map is essentially complete, ready to be labeled and drafted for reproduction.

The survey of Mud Glyph Cave was completed during four visits to the cave in early 1982. Jim Nicholls, who drafted the map of the cave from the survey data, concentrated primarily on sketching throughout the survey. The other participants—Bill Deane, Walter Merrill, and Art Cathers—split up the remaining survey chores.

In keeping with common archaeological practice, the angular measurements used to define the orientation of survey shots were made with a tripod-mounted Leitz Brunton compass and clinometer. A 30 meter Keson open-reel fiberglass tape, marked off in centimeters, was used for distance measurements. In two of the more difficult survey areas, a constricted side passage near the entrance and the extremely wet passage near the siphon in the back part of the cave, a hand-held Suunto KB-14/360-R compass and Suunto PM-5/360PC clinometer were substituted for the Brunton.

These latter instruments are widely used in cave surveys and have largely replaced the Brunton and various other instruments among cave mappers in many areas. Like the Brunton, they provide means for sighting on survey points and can be conveniently lighted for use underground. They are very well suited for work in tight passages, and they can be made relatively waterproof.

When used with a tripod, however, the Brunton has some major advantages because the instrument itself can be used as a temporary survey point, measuring the shot back to the previous point and then pivoting the instrument and shooting to the succeeding point. Alternatively, the Brunton can be set up directly over a previously measured point on the floor of a passage, measuring only the vertical distance between that point and the instrument before shooting to the succeeding point. Both procedures allow the surveyors to conveniently use survey points on the floor of the cave, largely eliminating the need for points marked on the walls of passages. Given the nature of the archaeological evidence at Mud Glyph Cave, this was a significant advantage, and the Brunton was the instrument of choice in most of the cave.

Measurements with all of the instruments were estimated to the nearest tenth of a degree, and the measuring tape was read to the nearest centimeter. The overall uncertainty in each shot was estimated to be on the order of a few tenths of a percent of the length of the shot, given the probable errors in each of the measurements.

Two survey loops in the rear section of the cave provided an opportunity to empirically check the accuracy of the survey. At the entrance to the glyph gallery, the survey split into two branches, one proceeding through the gallery and the other following the parallel stream passage. The branches rejoined at the end of the gallery, forming the larger loop, and were also connected through a window linking the two passages, to form a short loop. In each loop, data from both branches of the loop were used to calculate the position of the survey point where the branches rejoined.

In general, because the individual measurements made during a survey are not perfectly accurate, the two calculated positions for the final point will differ somewhat, and the calculated distance between them is the closure error for the loop. For the larger loop at Mud Glyph Cave, the closure error was 0.5 percent of the total survey distance around the loop. Results for the smaller loop were comparable, and in both cases the actual closure errors were small enough to ignore, since their effects on the map would be negligible.

The first survey trip to the cave was productive but largely uneventful. Before commencing work in the cave, a few survey shots were made to points of interest just outside the entrance, including the spring where the cave stream emerges. The survey was then slowly extended underground, following the stream in the back sections of the cave, and avoiding the glyph gallery entirely. Near the upstream end of the cave, the survey crew elected to stay relatively dry, diverting the survey into the large upper level room and leaving the stream crawl to the siphon for a future trip. In total, forty survey shots were measured in the cave, covering approximately 333 meters of passage.

The glyph gallery was surveyed on the second trip to the cave and a slight modification of the usual survey technique was necessary to accommodate the permanent reference markers, the datum points, along the length of the passage. Each marker was a one foot long piece of half inch by half inch aluminum L-beam, labeled with a meter number, and pressed into position in the mud floor. In order to leave the walls and untrampled areas on the floor of the passage undisturbed, the survey followed the trail established on earlier trips through the passage as closely as possible.

For each survey shot, the Brunton served as the initial point. A short length of the survey tape was stretched from the Brunton to a spot in the passage ahead, and held to that position. The numbered markers were then implanted in sequence, each one under the tape and a meter beyond the preceding one. The survey then proceeded as usual, using the base of the farthest meter marker as the next survey point and making the standard measurements. The Brunton was then moved into position above the new survey point. After measuring the vertical distance from the point to the Brunton, the process was repeated. At each end of the gallery, and at one area in between, the survey was tied into points established during the previous trip, forming the loops mentioned earlier. At the conclusion of the gallery survey, ninety-six markers had been placed in position, marking out a series of line segments along the length of the passage. One final shot was made in locating a separate marker in the stream passage, near the petroglyphs on the wall. In total, only thirteen survey shots were made, adding 103 meters to the survey.

The only problems encountered in this critical part of the survey stemmed from variations in the depth of the mud deposited on the gallery floor. Near the northeast end of the passage, the mud proved to be quite thick, and the markers were easily pressed into place, leaving only a few inches at the numbered end of each marker exposed above the floor. At the southwest end of the gallery, on the other hand, the deposit was generally too thin to support the markers. The situation was remedied with globs of mud collected from the stream passage, which were used to fashion small supporting mounds around the markers. The mounds dried quickly, and these markers proved to be reasonably permanent and highly visible.

The known extent of the cave grew during the third survey trip, with the discovery of a new section of passage. Such discoveries are relatively commonplace during cave surveys, but in this case the find was unexpected. The cave had been thoroughly explored prior to the beginning of the survey and seemed very unlikely to hold any surprises for the survey crew. The upstream end of the cave, in particular, had been carefully examined. Beyond the climb-up into the upper level room, the cave continued to the northwest for several meters as a crawlway, with the stream covering most of the floor, finally ending in a small room. Pooled-up stream water filled the bottom of the room, and the source of the water was found to be a small opening about a foot below the water surface on the left side of the room. Since the hole was too small to permit human passage, the cave effectively ended at this siphon, albeit temporarily.

On the third and supposedly final survey trip, Jim Nicholls and Art Cathers entered the cave, intending to mop up small leads near the entrance and to survey the short section of wet passage leading to the siphon using the Suunto instruments and mylar sheets for the sketch and notes. Upon arriving at the siphon, however, it was discovered that a great deal of silt had been washed out, probably during a recent period of heavy rainfall and flooding. A relatively large hole had been opened below water level and a separate softball-sized hole was open above the water, through which more passage could be seen.

After sitting down in the water and sticking his legs through the siphon verifying that his feet could be seen above water on the other side of the small hole and that the siphon was now large enough to easily fit through, Nicholls took a deep breath and proceeded to squirm through, surfacing in a low air-space on the other side. After reconnoitering briefly and finding some large passage, he returned to the siphon, and the survey was extended to a point on the far side by making a shot through the hole. Cathers then passed his carbide light through the hole and dove through, guided by the light from Nicholls's electric headlamp, which was reasonably waterproof.

Approximately 70 meters of cave passage was found beyond the siphon, most of it large enough to walk through. The cave ended at an apparently impenetrable breakdown plug, with the stream issuing from several small openings in the breakdown. No sign of previous human visitation was noted. After exploring briefly, both cavers concluded that they were too cold to finish the survey, and they quickly left the cave.

Bill Deane, easily convinced that a return trip would be necessary to produce a complete map, joined Nicholls and Cathers on the fourth and final survey trip. All three came prepared to deal with the siphon by wearing wetsuits and electric headlamps. After passing note pads and the Suunto instruments through the hole above the siphon, the survey began at the last point set on the previous trip, covering the new section of cave in seven shots. After a thorough check for signs of previous visitation and for any routes that might lead into additional passage, the party left the cave, with the survey now completed.

It should be noted that siphon diving is not generally a recommended practice and is considerably more dangerous than caving in general. The situation at Mud Glyph Cave was extremely favorable for a successful dive, with a short, relatively open section of submerged passage, and with air-filled passage clearly visible beyond the siphon. Under most circumstances, siphons are best left to those with the experience and equipment necessary to dive safely; there is no point in drowning while trying to see a few additional meters of cave passage.

Reduction of the survey data was begun shortly after the first trip, and most of the map had been completed well before the last two survey trips took place. Since nothing of archaeological interest had been found beyond the crawl leading to the siphon, it was decided that that portion of the cave could be omitted from the maps intended for research use. In fact, because of the more or less linear layout of the cave, it was necessary to portray the cave in two sections in order to show sufficient passage detail without producing an unmanageably long map. The final project map, drawn at a scale of 5 meters to the inch, showed roughly 450 meters of the cave, with the front part of the cave, from the entrance to the glyph gallery, at the top of the map, and the remainder of the cave below. Figure 1 was redrawn from that map, omitting some detail but adding on the back section of the cave.

A second map, showing only the glyph gallery and adjacent passages, was drawn at a scale of 4 meters to the inch, and included the locations of several glyphs and other features of interest and considerably more passage detail. Figure 2 was redrawn from this second map, again omitting detail, due to the reduction in size.

Both maps were reproduced in blue line form, and copies were provided to all research participants.

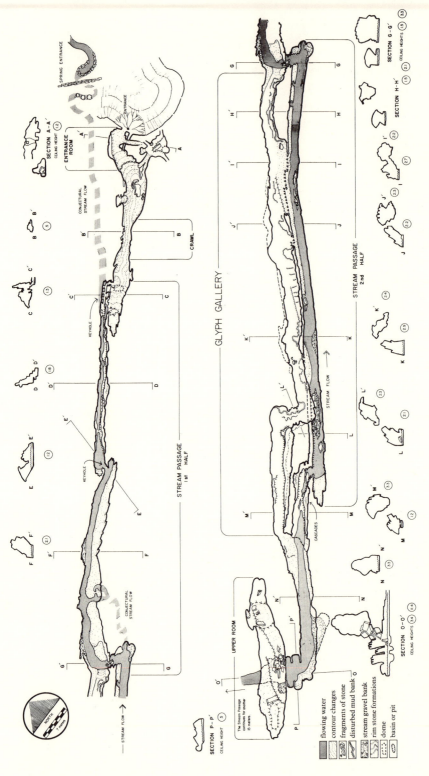

FIGURE 1: MAP OF MUD GLYPH CAVE.

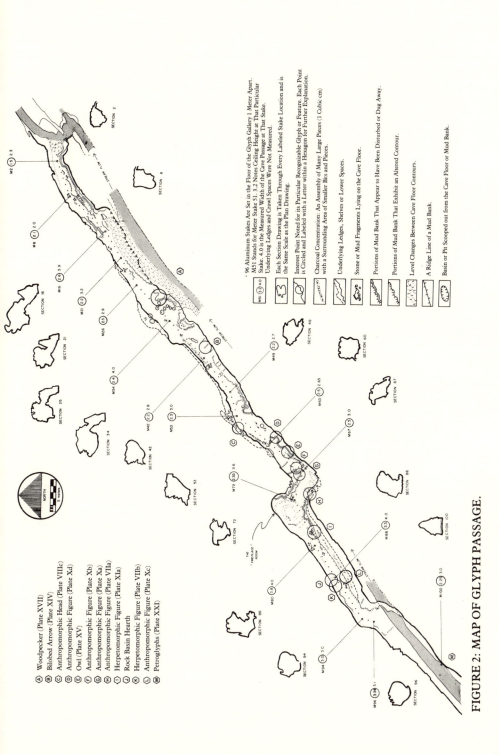

(A) Woodpecker (Plate XVII)
(B) Bilobed Arrow (Plate XIV)
(C) Anthropomorphic Head (Plate VIIIc)
(D) Anthropomorphic Figure (Plate Xd)
(E) Owl (Plate XV)
(F) Anthropomorphic Figure (Plate Xb)
(G) Anthropomorphic Figure (Plate Xa)
(H) Anthropomorphic Figure (Plate VIIa)
(I) Herpetomorphic Figure (Plate XIa)
(J) Rock Basin Hearth
(K) Herpetomorphic Figure (Plate VIIb)
(L) Anthropomorphic Figure (Plate Xc)
(M) Petroglyphs (Plate XXI)

'96 Aluminum Stakes Are Set in the Floor of the Glyph Gallery 1 Meter Apart. M51 Stands for Meter Stake 51. 3.2 Notes Ceiling Height at That Particular Stake. 4.0 is the Measured Width of the Cave Passage at That Stake. Underlying Ledges and Crawl Spaces Were Not Measured.

Each Section Drawing is Taken Through Every Labeled Stake Location and is the Same Scale as the Plan Drawing.

Interest Point Noted for its Particular Recognizable Glyph or Feature. Each Point is Circled and Labeled with a Letter within a Hexagon for Further Explanation.

Charcoal Concentration: An Assembly of Many Large Pieces (1 Cubic cm) with a Surrounding Area of Smaller Bits and Pieces.

Underlying Ledges, Shelves or Lower Spaces.

Stone or Mud Fragments Lying on the Cave Floor.

Portions of Mud Bank That Appear to Have Been Disturbed or Dug Away.

Portions of Mud Bank That Exhibit an Altered Contour.

Level Changes Between Cave Floor Contours.

A Ridge Line of a Mud Bank.

Basin or Pit Scooped out from the Cave Floor or Mud Bank.

FIGURE 2: MAP OF GLYPH PASSAGE.

III

Recording the Glyphs:
Photo Documentation
of Mud Glyph Cave

The photo documentation of Mud Glyph Cave offered a unique challenge. It was not just a simple matter of walking into the cave with a camera.

There were seven photo trips to the cave, each lasting from eight to twelve hours underground. A total of 323 color slides and 323 black and white negatives was taken to document the glyphs. All 323 black and white negatives were enlarged to 8″ x 10″ prints by David Luttrell, a Knoxville photographer. An additional 150 color slides were taken, showing scenic views of the cave and of the researchers at work.

Caves thrust the photographer into an environment that is at once hostile and intriguing. The photographer and his equipment are surrounded by jagged rocks, squishy mud, dripping water, and a subtle, numbing cold. Mud Glyph Cave has an ambient temperature of 54 degrees F with near 100 percent humidity.

To protect my camera gear, I used Army surplus .50 calibre ammo cases lined with foam sponge. Holes are cut into the foam to form-fit the cameras, lenses and flash units. This helps insulate the equipment from jarring bumps as the ammo case is dragged through crawlways. An ammo case is nominally waterproof. If the case is accidently dropped into a few feet of water and retrieved immediately, the camera gear should survive.

The choice of which brand of camera equipment to use is a purely subjective matter. All gear should be sturdy enough to survive the bruising trip through the cave. In general, wide angle and normal focal length lenses are best due to the confining nature of cave passages. I used Nikon 35 mm cameras and Nikkor lenses of 24 mm, 35 mm, 50 mm, and 55 mm focal length for the photography in Mud Glyph Cave.

The photo documentation included a complete series of both color and black and white shots. I chose Kodak Kodachrome 64 (ASA 64) color slide film because it is the finest grained medium speed color film available. I chose Kodak Tri-X Pan black and white negative film (ASA 400) because it is a high speed film with very fine grain. I had used both these films before for cave photography with good results.

One of the most important tools of a cave photographer is a good sturdy tripod. It is not necessary to use a tripod for every shot. The tripod, however, can help the photographer to compose his photograph and free his hands for other tasks such as firing a flash. A tripod-mounted camera produces a sharp, jiggle free image. A reliable locking cable-shutter release is also a must.

To speed up the photography, two 35 mm cameras (one for color, one for black and white) were mounted on the tripod using a special bracket. This way, both cameras could be opened and exposed to the same flash simultaneously. This also saved the number of flash bulbs used.

Because caves are dark (except for the personal light sources brought in by the cavers), the shutter can be left open, using the bulb setting with no fear of light fogging the film.

Folding-fan flash units using M–3B flashbulbs were selected as the light source for the Mud Glyph Cave photographs. In retrospect, since over 500 bulbs were used to photograph the cave, it would have been easier, less expensive, and just as effective to have used electronic flash units. In the case of electronic flash units, it is always a good idea to bring along spare batteries.

The walls of Mud Glyph Cave are covered with a layer of dark mud. These dark walls reflect very little light and make it difficult to estimate the proper exposures to use in the cave. A series of test photographs was taken in the cave with the M–3B bulbs to determine the exact guide numbers for the Kodachrome 64 and Tri-X Pan films.

To simplify the photography, I decided to always have the M–3B bulb held six feet away from the subject. This allowed the same pair of f-stops (f/9.5 for the color and f/22 for the black and white) to be used for all of the photographs. A simple knotted string was used to measure the six-foot distance.

The exact angle of the flash bulb proved to be very crucial in illuminating important details of the glyphs. Because of the dark walls and the very faint tracings on many of the glyphs, a direct head-on light source could "wash out" details. We discovered through trial and error that side lighting with a single flash produced the best images.

For each shot, an assistant would first illuminate the subject with a coal miner's electric lamp from several different angles. Meanwhile, I would view the subject through the camera to determine the best lighting angle and flash position for each individual photograph.

A scale painted red and white in 2.5 centimeter increments appears in each photograph (in this book, black and white is reproduced). I would position this scale so that it would not cover up any important feature of the glyphs.

On each trip, two assistants were necessary to help with the photography. The individuals involved were Art Cathers, Jim Nicholls, Melvin Whaley, and John Yust. Their efforts contributed greatly to the success of the photography.

One assistant was responsible for the lighting of the pictures. This involved measuring the six-foot distance, determining lighting angle with the coal miner's lamp, and firing the flash each time.

The other assistant recorded the film frame number, a brief description of the subject glyph(s), and the position of the glyph with regard to survey stakes that had been placed along the center of the passage at one meter intervals when the cave was mapped.

Since many of the glyphs are similar, this seemingly excruciating record taking was absolute necessary. The photographs of the glyphs were cataloged by right or left wall and meter stake number. This allowed the researcher back in the office to orient adjacent and overlapping photographs in order to reconstruct the "mural" effect seen on the cave walls.

The mud glyphs will not last forever. A change in the cave environment (such as a massive flood), an accidental mistake by a well-meaning visitor, or the purposeful destruction of these features by a vandal can erase in seconds what has survived for centuries. The color slides, which are now carefully stored at the University of Tennessee at Knoxville, will only last for twenty to thirty years. The black and white negatives should last for 100 to 200 years. Eventually scientific progress may allow the photographs to be stored as an electronic signal in a computer. Thousands of years from now, we hope, long after the glyphs have faded into dust, our photo documentation will still survive.

CHARLES H. FAULKNER
ROBERT STUCKENRATH
GARY D. CRITES

IV

Dating the Glyphs:
Charcoal and Stoke Marks

An archaeological survey of the cave floor was conducted to recover any artifacts that may have been lost by the aboriginal cavers, to locate any disturbances these people might have made in the cave floor, and to collect charcoal samples for radiocarbon dating. No lithic, ceramic, or bone artifacts were recovered in this survey. The entire floor of the glyph passage was carefully searched for such artifacts, especially under the ledges where the floor was relatively undisturbed. Besides the animal scat described in Chapter V, the only non-charred organic remain preserved on the floor was a small fragment of eastern redcedar wood (*Juniperus virginiana* L.) recovered at 80 L. While this could be the remains of a wooden artifact, it is just as likely this fragment is an unburned piece of firewood. While the extreme dampness of this cave has helped preserve the fragile glyphs, it has caused the almost complete deterioration of any non-charred plant remains that might have been left by the aboriginal visitors.

Indirect evidence of a possible wooden artifact carried into the cave is the impression of a bark-covered club or pole at 50L. The shaft of this wooden implement was struck against the wall leaving a clear impression of the bark which has been identified as hickory (*Carya* spp.) A number of small round depressions (5.5–7.5 cm in diameter) were also found on other areas of the walls, sometimes mutilating underlying glyphs. A cast of these impressions indicates these marks were caused by the blunt butt end of a wooden club or pole.

While the entrance passage appears to have been substantially modified since the first humans set foot into Mud Glyph Cave, the glyph gallery may not have been disturbed until it was entered by modern cavers in 1980. Unfortunately, these recent visits have erased any vestiges of aboriginal footprints on the floor of this passage. The extensive use of the footholds cut into the mud bank, providing access to the glyph gallery from the lower stream passage, makes it impossible to determine if these depressions were dug by the Indians. Our impression, however, is that they date from the aboriginal utilization of the cave.

The only clearly demarcated aboriginal footprint is located at 55–56R. This is the bare print of an adult individual who put his or her right foot against the wall to apparently push herself or himself through this narrow section of the passage. During a later visit an aboriginal artist encircled the heel impression creating the shell of a realistically executed turtle, or what Jon Muller believes may also be a stylized human face (see Chapter VI). This zoomorphic glyph is paired with a human mask glyph; close examination of the trailed lines in these drawings indicated they were produced with the same implement, probably a piece of cane.

Two types of features were found on the glyph gallery floor. One type was a shallow depression; several of these features occurred near the entrance to the glyph gallery. One of these depressions at 90–91R at the end of the passage appears to have been scooped out with the soil mounded around the perimeter. The basin measured 80 cm in diameter. A 15 cm x 50 cm profile trench excavated to the center of this feature revealed an upper stratum of red brown clay and lower stratum of silty brown clay resting on a weathered rock base. The contact surface between the red brown and light brown silty clays appeared to be burned in places. This appears to be a prepared hearth, but apparently was not used more than once.

The other feature type was a hearth within a small natural rock basin at 86R. The irregular rim measures about 11 cm in diameter. A thick ash fill and heavily burned rim indicate the basin was probably used several times. A sample of hearth charcoal was identified as hickory (*Carya* spp.), ash (*Fraxinus* spp.), and honey locust (*Gleditsia triacanthos*). Since wood is not washed into the cave by the stream, a mixture of firewood was obviously carried into the cave by the aboriginal artists. One might also infer from this behavior that they entered the cave with the intent of building small fires to illuminate their work; i.e. the drawings are not the result of spur-of-the-moment "doodling" by occasional cave explorers.

Other concentrations of charcoal presumed to be small ad hoc hearths were reported by the 1980 discoverers of the glyph gallery, especially in the widest and highest portion of this passage called the "fireplace room." Unfortunately, this area had been so extensively trampled by visitors before the survey team entered the cave that none of these purported hearths could be discerned, although a considerable amount of charcoal is still present on the floor in this locality.

While other small, single event hearths may have been present on the glyph passage floor, the presence of only one intensively used hearth at meter stake 86 suggests most of the illumination of the passage was by torch light. Experiments with cane (*Arundinaria*) torches in Salts Cave (Watson 1969:33–36), consisting of a three-stalk bundle 3 feet long, indicated such a torch would burn for almost three quarters of an hour. Because the modern trip back to the glyph gallery takes between ten and fifteen minutes, an artist could theoretically reach the glyph gallery, decorate a small section of the wall, and leave the cave with one torch. This is assuming, of course, that the Indians used the same passages we do today. While the lower stream passage occasionally floods during periods of heavy rainfall, the glyph gallery would have been easily accessible during most of the year with a minimum of stooping and crawling. It is believed the Indians entered the cave by the same entrance as we do today, although the stream passage may have been more open at that time. The flooding in this lower passage, however, has destroyed all traces of such prehistoric traffic there.

All of the above evidence suggests that the aboriginal visits were of short duration. This is also supported by the fact that no stone or bone tools or ceramic containers were carried into the cave. It may also be significant that the only area where hearths are found is adjacent to the high ledge on which the large drawings were made. This might indicate stronger illumination was needed for a more extended period of time as these more inaccessible areas were decorated.

Another factor affecting the length of stay in this passage might be smoke from these fires. Modern experiments with cane torches in Salts Cave indicate a few minutes of torch use causes a permanent haze in the passage and reddened eyes and runny noses on the part of the cavers (Watson 1969:36). Small fires would have caused an even more dense smoke in the gallery. The clay walls and glyphs have a black film over them which is probably the sooty residue from such smoke. Modern marks on the clay show a bright orange under-surface.

The most common remain left by the aboriginal visitors on the passage floor is charcoal, this material being particularly concentrated along the natural "trail" down the center of the passage, under the rock ledges, and beneath certain clusters of glyphs. Charcoal samples removed from the floor consist primarily of cane (*Arundinaria gigantea*), although some pine charcoal (*Pinus* sp.) is also present. Both cane and weed stalks (*Gerardia* or *Solidago*) were used as torch material in Salts and Mammoth caves (Watson 1969:33; 1974:183–92). It is assumed that most, if not all, of the cane charcoal on the floor of the glyph gallery is from torches. While it is likely that some of the pine charcoal is from small surface fires, pine torches could also have been used by the aboriginal visitors. Scattered pine charcoal in Williams Cave in Bath County, Virginia, has recently been dated to the eleventh century A.D. (Faulkner 1986).

Despite the considerable amount of charcoal on the gallery floor, few stoke or smudge marks are present on the walls. A good stoke mark can be seen on a section of exposed limestone at 78L beneath the high ledge. This could be further evidence that the aboriginal visits were of short duration, although a more speculative interpretation might be that the Indians deliberately avoided touching the clay-covered walls with their torches. What appears to be another light stoke mark on the low ceiling near the siphon is also informative. If the floodwaters in this lower passage never reached the ceiling in this area of the stream passage and this smudge is indeed from a cane torch, it indicates the aboriginal cavers explored the stream passage as far as the siphon.

Eight charcoal samples collected in the glyph passage were dated at the Radiation Biology Laboratory, Smithsonian Institution. These included five samples of cane and one of pine charcoal from the floor in areas of more intensive activity or under specific glyphs, and two samples of hard and soft woods from the rock basin hearth. The Radiation Biology Laboratory sent carbon dioxide samples generated from the cane samples to Geochron Laboratories for C–13/C–12 analysis to provide corrections to the age calculations. These uncalibrated C–14 dates are found on Table 1.

The date of A.D. 465 ± 60 years for the cane charcoal under the mask and turtle glyphs suggests Mud Glyph Cave might have been entered as

TABLE 1. Radiocarbon Dates from Mud Glyph Cave.

Sample No.	Material	Provenience	Uncalibrated Date C-13 Corrected
SI-5468	cane charcoal	beneath mask and turtle glyphs; meter 56-57	A.D. 465 ± 60
SI-5098B	soft and hard-wood charcoal	basin hearth; meter 86	A.D. 1155 ± 60
SI-5470	pine charcoal	fireplace room meter 70-71	A.D. 1200 ± 45
SI-5098A	soft and hard-wood charcoal	basin hearth; meter 86	A.D. 1235 ± 60·
SI-5469	cane charcoal	near mask and turtle glyphs; meter 56-57	A.D. 1315 ± 50
SI-5471	cane charcoal	under high ledge; meter 78-79	A.D. 1335 ± 60
SI-5472	cane charcoal	under high ledge; meter 79-80	A.D. 1605 ± 65
SI-5473	cane charcoal	near end of glyph passage; meter 90-91	A.D. 1760 ± 80

(after Faulkner, Deane, and Earnest 1984)

early as the Middle Woodland Period. Since these two glyphs are drawn in the same manner as the others on the walls and sample SI-5469 from the floor near-by dates A.D. 1315 ± 50 years, the occurrence of early dated charcoal beneath them is undoubtedly fortuitous. Middle Woodland dates from Wyandotte Cave in Indiana and Big Bone Cave in Middle Tennessee (see Chapter VIII) indicate the Indians entered caves during this time period in eastern North America. Nevertheless, the fact that seven of the eight dates fall between the twelfth and eighteenth centuries indicates sample SI-5468 might be in error. There is no question, however, that the most intensive visitation took place in the Mississippian

TABLE 2. Mississippian Radiocarbon Dates from
the Eastern Tennessee Valley

Sample No.	Site	Uncalibrated Date	Source
HIWASSEE ISLAND CULTURE			
GX-2594	Leuty Mound (40RH6)	A.D. 1100 ± 100	(Schroedl 1978)
GX-4212	Martin Farm (40MR20)	A.D. 1160 ± 130	(Schroedl 1978)
M-729	Bowman Farm (40CP2)	A.D. 1190 ± 150	(Crane and Griffin 1961)
GX-4211	Martin Farm (40MR20)	A.D. 1195 ± 140	(Schroedl 1978)
GX-6077	Toqua (40MR6)	A.D. 1215 ± 130	(Polhemus 1986)
GX-1572	Mayfield Mound (40MR27)	A.D. 1250 ± 95	(Salo 1969)
M-731	DeArmond	A.D. 1280 ± 150	(Crane and Griffin 1961)
DALLAS CULTURE			
GX-6075	Toqua (40MR6)	A.D. 1345 ± 120	(Polhemus 1986)
GX-4205	Toqua (40MR6)	A.D. 1480 ± 130	(Polhemus 1986)
GX-6076	Toqua (40MR6)	A.D. 1550 ± 125	(Polhemus 1986)
GX-4207	Toqua (40MR6)	A.D. 1620 ± 120	(Polhemus 1986)
M-582	Williams Island (78-81HA60)	A.D. 1620 ± 75	(Griffin 1963)
GX-6074	Toqua (40MR6)	A.D. 1635 ± 120	(Polhemus 1986)
GX-4206	Toqua (40MR6)	A.D. 1670 ± 105	(Polhemus 1986)

cultural period about 700 years ago. Five of these dates range between the twelfth and fourteenth centuries and the mean of the seven later dates is A.D. 1248.

This date range from the twelfth and eighteenth centuries indicates the glyphs could have been drawn by both the Hiwassee Island and Dallas cultures. Village sites of both of these cultures are widespread in the eastern Tennessee Valley, although none have been found in the immediate area of Mud Glyph Cave. For comparative purposes, uncorrected Hiwassee Island and Dallas culture radiocarbon dates are found on Table 2. The mean radiocarbon date of A.D. 1248 suggests many of the glyphs were drawn by either late Hiwassee Island or early Dallas artists. Because there is growing archaeological evidence that Hiwassee Island is ancestral to later Dallas cultural development, an attempt to separate these two manifestations in the thirteenth century is probably futile. Nevertheless, if we must assign this art to one culture or another, the fact that several glyph motifs are the same as those found on Dallas stone, ceramic, and shell artifacts indicates an association with the Dallas people.

RONALD C. WILSON

V

Scratches on the Walls:
Other Cave Visitors

Caves are attractive places of refuge for many animals. Most are small arthropods such as beetles, crickets, and spiders. These residents are generally small and escape the notice of most cave visitors. Crayfish often occur in cave streams and are the largest invertebrate animals routinely encountered in caves. Bats are among the best known vertebrate animals associated with caves. A few species form maternity colonies in caves, and many bat species use caves as hibernacula. Other animals that routinely use caves as den sites include rodents such as woodrats and carnivores such as foxes, bears, bobcats, and raccoons. Caves provide such mammals with shelter from extreme weather and protection from predators.

Mud Glyph Cave is typical of eastern North American caves in its utilization by a variety of animals. Most concentrate their use of the cave in the twilight zone, but a few venture into the deepest recesses of the cave. During both historic and prehistoric periods throughout the world, human and non-human animals have sought the relatively stable environments of caves for many of their activities. The same stability of environment that was the original attractant also preserved evidence of these activities for our interpretation. This chapter deals with the evidence of non-human vertebrate biological activity in the glyph gallery passage of the cave.

Drawings, scratches, and other marks cover almost all mud covered surfaces in the glyph gallery. Many are clearly representations of people, animals, or symbolic motifs such as the weeping eye. Others are not so easily interpreted. A series of vertical scratch marks that occurs along the base of the gallery walls is of special interest to me as a biologist. These scratches are generally confined to the lower eighteen inches of the cave walls. They do not occur on higher ledges that are covered with elaborate artistic glyphs but do occur along portions of the stream passage as well as in the glyph gallery. The scratches occur in parallel groups of four. Some areas are so heavily scratched that individual groupings are difficult to discern. In some cases they resemble the marks of the clearly cultural glyphs. In other cases they are associated with impressions of toe and metapodial pads. On close examination the shape of the individual scratches is unlike the marks of the glyphs. They are clearly the result of living animal activity.

No complete animal tracks are preserved, limiting the usefulness of standard references on animal tracks (e.g. Murie 1975). In most caves that contain animal tracks, however, both complete and incomplete footprints occur. If a single species is responsible for all marks in a particular passage, the possibility exists for a complete record of the spectrum of markings that species may produce. Such records are available on the floors and walls of many Southeastern caves.

Jaguar Cave (Fentress County, Tennessee) contains tracks and scratches left by opossum, bobcat, otter(?), bats, and Pleistocene jaguars (Robbins, Wilson, and Watson 1981). Tracks and clawmarks of black bear are abundantly preserved in Precinct Eleven Cave (Rockcastle County, Kentucky) and in Cumberland Caverns (Warren County, Tennessee). Cave Hollow Cave (Lee County, Kentucky) contains tracks of bobcat and raccoon. Raccoon tracks are also preserved in many other regional caves including Rider's Mill and Cooch Webb Caves (Hart County, Kentucky), and Caney Branch Cave (Clinton County, Kentucky). The scratches and partial tracks of Mud Glyph Cave were made by raccoons (*Procyon lotor*).

Raccoons occur throughout North and Central America except in portions of the Rocky Mountain region and the Great Basin (Hall and Kelson 1959). In Tennessee the breeding season begins in late January and lasts through the summer. Young are born from early April to October (Linzey and Linzey 1971; Golley 1962). They are omnivorous but prefer foraging along the margins of streams and ponds (Barbour and Davis 1974).

Hollow trees are preferred den sites, but caves and rock crevices are also routinely used (Golley 1962; Doutt et al. 1967).

Berner and Gysel (1967) studied the use of dens by raccoons. Ground burrows were used most from October to mid-February and least in April, May, and June. Temperature was especially important during the winter as a factor determining den activity. During the cold of mid-winter, raccoons often stayed in the dens. During mid-July through September warm weather and abundant food made the use of the dens less urgent. Activity at the entrance to dens peaked at 4–5 A.M. and 7–8 P.M., illustrating the nocturnal habits of the species.

Based on these observations, the raccoon activity at Mud Glyph Cave most likely occurred between late fall and early spring when the cave was being used as shelter. The primary use of the cave by raccoons was for shelter, although it is possible that they occasionally found food in the form of crayfish from the cave stream or, less frequently, a bat. Careful study of the glyphs indicates that most claw marks predate the glyphs and a few postdate the glyphs. The raccoon's scratch marks were left as an inadvertent by-product of their inquisitive explorations. Two animal scats collected from the cave by Howard Earnest in March 1982 were highly decomposed but are referred to as raccoon. One specimen contained fragments of bat bones and beetle parts. The other was composed entirely of pumpkin seeds. These scat samples confirm that raccoons still frequent the cave and further support the suggestion that raccoon use was most likely during the fall and winter. Similar seasonal predation on bats by raccoons in Wyandotte Cave, Indiana, is well documented by Munson and Keith (1984). Raccoon activity in the cave is independent of the human activity that produced the exceptional mud glyph art. The stable cave environment preserved evidence of both human and animal activity indiscriminately.

Serpents and Dancers: Art of the Mud Glyph Cave

THE CONTEXT

The Mud Glyph Cave is located in a somewhat isolated area of eastern Tennesseee away from the larger centers of its own times. Thus interpretation of the place of this cave and its graphic representations in Eastern prehistory is complex. The cave itself has over 100 meters of gallery with designs incised into the clay covering left on both sides of the cave walls (see Chapter II). At first glance, the impression given by the cave art is somewhat disappointing—the execution of the various representations appears crude compared to those seen on "valuable" and more durable items such as shell gorgets and copper plates. To some extent this is a reflection of the apparently hurried execution of the "glyphs" (a term used for convenience, merely indicating that these are a "sculptured mark or symbol" [OED]). As in any case, "crudeness" of drawings may be explained by many factors. It is possible that the local style of representation was simply less "formal" than those of other localities. It is also possible that the representations were done by persons who were not skilled or knowledgeable about the art styles of their own society or that the makers were members of groups that were marginal in their relationships to the main Tennesseee Valley art styles. Careful examination of these drawings, however, shows considerable consistency in execution; and close study suggests that drawings are "cursive" rather than "crude." Whatever the situation, the representations are not easily fit into the more-or-less established art styles of the region as defined for more durable materials (Kneberg 1959; Muller 1966a, in press; Phillips and Brown 1975–82). There are many similarities in themes and motifs, but the similarities do not extend to specific stylistic features. Although the Mud Glyph cave materials are very unusual in the context of surviving Southeastern Late Prehistoric art, it has to be remembered that the glyphs found in the cave may be representative of a long-lived artistic tradition which has left us few other archaeological traces. Historic records from two or three hundred years later describe elaborate wall decoration (e.g., Bartram 1928 [1791]:361), and it is likely that similar drawings formed a part of the decoration of structures throughout much of the Southeast. The isolated and hidden character of the Mud Glyph Cave means that it cannot be taken to be representative of art forms that were more publicly displayed, but at least the style of representation may be one that was more likely used than the few surviving examples indicate.

THE MEANING OF THE ART

> Lowell always said that the regularity
> of the canals [on Mars] was an un-
> mistakable sign that they were of in-
> telligent origin. This is certainly true.
> The only unresolved question was
> which side of the telescope the in-
> telligence was on.
> Sagan 1980:110

Unlike the supposed Martian canals, there can be no doubt of the human origin of most of the "glyphs" scratched into the soft clay coating of the walls of the cave. Nonetheless, the interpretation of the cave wall markings is very much subject to the "Martian problem," as the combinations of line and form are so convoluted and overlaid as to make identification of motifs and themes subject to substantial interpretive difficulties. Much of the wall is a veritable palimpsest, with earlier figures erased by wiping, or simply overdrawn without erasure. Since the style of the drawings and the techniques used to make the lines remained fairly uniform through time, it is often very difficult to tell what lines go with which other lines. In addition, there is evidence that natural cracks, protrusions, and other non-artificial features of the cave wall were incorporated as parts of designs.

The first modern explorers of the Mud Glyph Cave gave a series of nicknames to various glyphs which illustrate the capability to see order in these prehistoric drawings. It is, of course, clear that a "jogger" or "Pogo" are not what was intended by the prehistoric artist. These are merely convenient labels for the representations in the cave but demonstrate the truism that modern viewers who struggle to identify the character of the cave representations cannot help but see the materials in terms of their own perceptions and experience. Thus most of the human representations were first seen as "stick" figures, but examination of these figures shows that there is much more to the design than was realized. The danger, of course, is that too much can be read into these representations, that the interpretation can put the intelligence on the wrong end of the telescope, as it were. It will be noted that the interpretations of motifs here differ in a number of respects from the preliminary interpretations in the initial published account (Faulkner, Deane, and Earnest 1984). This should not be taken as criticism of that early effort, with which I was in substantial agreement at the time it was written, but merely an indication of the results of more intensive and continuing reassessment of the drawings. What follows is a preliminary analysis of the Mud Glyph Cave drawings, but it cannot be emphasized too strongly that it would be a serious mistake to attempt to "interpret" these themes and motifs in terms of Southeastern myth and religion as they were in the Historic period. This is especially true given the difficulty in identifying the intended subject of many representations. Moreover, the glyphs were drawn more than 200 years earlier than our earliest, fragmentary historic records on the region. In addition, the bulk of our historic data on this region is from a period following tremendous dislocation, depopulation, and acculturation as a result of European political interference and disease. The time and social change is roughly the same as the time between the American Revolution and World War II! It is not that the historical cases can tell us nothing about the earlier, but extreme caution has to be exercised. I will make some suggestions about possible relationships of some of the motifs to historic mythology at the end of the paper, but these are not meant as guides to the semantics of the art.

Before any reasonable, scientifically testable suggestions can be made about the "meaning" of the glyphs, there will have to be a great deal of detailed study of the morphology and structure of the drawings. The drawings are so complicated and there is so much intertwining of elements, that efforts which start from the "meaning" are doomed to be futile and logically circular. No end of just-so stories can be constructed on an ad hoc basis to "explain" the elements, but any effort to learn about the past, rather than about our own feelings, has to start from careful contextual analysis as well as from a full understanding of the medium and the style.

The access to the cave was probably as difficult in prehistoric times as it is today. Although an easy cave to enter by "caver" standards, it is not so easy for a novice. There is only a short crawl, but the some 120 m to the gallery with clay covered walls must have been fairly exciting for people who may well have believed the caves were the homes of monsters of the underworld (see, for example, the historic beliefs of the Cherokee as described in Hudson (1976:167–68). Experiments with cane torches indicate that these were quite efficient light sources with a "pleasant, warm illumination" that make a cave "almost cozy, not mysterious or menacing" (Watson 1969:60), but modern torch bearers have different attitudes about the dangers of caves than did the aboriginal explorers. Torch light would have made the cave and glyphs appear different than would the high contrast of electric or carbide light. In addition, the flickering of a torch would have provided an element of animation that is lacking in steady light.

Every indication supports the argument that the Mud Glyph Cave was entered for purposes that are directly reflected in the drawings. There is no evidence for food preparation or habitation in the cave. No artifacts, other than the evidence of torches and fires, were found (see Chapter IV). The nature of the human use of the cave has to remain a mystery wrapped in an enigma, to some degree. It is quite possible that the action of drawing the glyphs was more important than the end product, but whether these actions constituted a ritual act like sand painting in the Southwest cannot be determined by presently available methods. The lack of variability in the representations also suggests that the cave art was the work of a relatively small number of individuals, although the cave may have been entered at other times. The presence of several good horizon markers from the "Southern Cult" period makes it likely that the majority of the cave art was done in the middle to late part of the thirteenth century. This corresponds well with the radiocarbon dates of A.D. 1315 ± 50, A.D. 1200 ± 45, and A.D. 1335 ± 60 (all C–4 dates are uncorrected). No art materials can be identified which might be from the time of the earliest date from the cave, A.D. 465 ± 60. Some of the large snake-like elements in the cave could, on thematic grounds, perhaps be forced into the A.D. 1605 ± 65 or A.D. 1760 ± 80 times, but internal evidence such as overlap of lines makes this very unlikely, given the basic uniformity of style in the cave.

THE GLYPHS

The analysis of the graphic materials in the cave must proceed on at least four different levels: 1) execution, 2) elements, 3) motifs and themes, and 4) structure. The difficulties increase for each succeeding level, but a complete analysis of each level is only possible if the other levels are taken into account. Of course the form of analysis is dependent upon the problems being studied by the analyst. The immediate problem here is the stylistic affiliation of this material with other areas in the Southeastern United States. Although this is a relatively low level problem, it is non-trivial in that the stylistic affiliations will play an important part in warranting hypotheses on the social and ideological implications of the art. As a "sealed deposit" the cave art at this site also presents us with a relatively discrete and complete segment of prehistoric behavior which is as worthy of study in its own right as any other part of human behavior, be it kinship or butchering practices. The Mud Glyph Cave provides us with a rare opportunity for isolating very "fine-grained," short-term human behavior in an archaeological context. For this reason, the scientific potential of the cave is very great, but the study of the cave will require very great effort and attention to detail. What follows is only a preliminary outline. In the following sections, each level of analysis is briefly discussed, but a detailed structural analysis along the lines developed for other Southeastern art (such as Muller 1966b, 1971, 1973, 1977, 1979) was not possible now. Even though the presentations are only preliminary, the necessary steps for future work are fairly clear. The following discussion also provides a motif and thematic inventory and catalog of the Mud Glyph Cave art.

EXECUTION

All representations in the Mud Glyph Cave were executed by drawing on walls of the cave with cane, sticks, or the human hand. In almost all cases, the drawings were done on damp clay lining the walls of the cave, but in a few places the relatively soft rock of exposed walls was incised. The clay was (and is) quite plastic and glyphs could be destroyed by a simple wipe of the hand. The plasticity of the medium is such that there is little difference in appearance between the ancient cave drawings and a few imprints left by "spelunkers" who entered the cave shortly after its discovery. The major damage to the glyphs, however, is either aboriginal or from natural causes. In some cases, the clay covering has dried enough to crack off from the underlying rock. More commonly, the various drawings in the cave were purposely defaced by smearing or stamping on the surface of the clay. The evidence suggests that this defacement was contemporary with the drawings and was probably a common termination of episodes of use of the cave. The most common damage to the drawings, however, is the obliteration of areas of the "mud" walls by raccoon scratching (see Chapter V).

ELEMENTS AND LINE MODES

The incised lines themselves are of several basic forms. The most common is a broad, shallow line, usually with faint longitudinal striations (probably from the end of a stick of cane). These lines are from about 7 to 20 mm wide and 4 to 5 mm deep. Some lines are smooth in bottom contour and could have been done with a smooth tool or with the tip of the finger. Other lines are narrow and V-shaped in profile. These were done with a sharp tool, something like a splinter of cane, or perhaps a knife blade of some kind. Although different kinds of lines overlie each other, there is no clear sequence in which, for example, broad-line incising was earlier and narrow, later. On the contrary, there is evidence of each kind of line being obliterated by the others.

There are other kinds of markings on the walls. In a number of areas, something, perhaps a hand, has been wiped over the surface of drawings in order to obliterate or obscure them. There are instances of what appears to be grabbing of the clay and pulling it away or down from the wall. No actual finger prints with identifiable whorl markings were observed by me, however. There is at least one clear foot impression, even including separate impressions of the toes, but I did not see any indications of prints that could be used to identify individuals. There are some possible sandal impressions, although the most identifiable foot impression is of the naked foot. Other impressions may be of knees or heels dug into the surface of the clay while working on higher surfaces. Many of the drawings have been obliterated by striking the clay with a round object, possibly the butt of a wooden torch or club. Charcoal fragments in the cave indicate that some wood has been brought into the cave for fuel, although cane torch fragments are common (see earlier chapters). In one area, a 5 cm diameter piece of hickory was struck sideways against the clay, leaving a clear impression of its bark. In a few cases, designs, some of which appear to have been the heads of human figures, were cut out from the wall following the outline of that portion of the drawing. There are also a large number of cane or small stick impressions where the end of the cane has been poked into the surface. At least some of these have been used as parts of larger designs (as "eyes," for example). Other, more irregular jabs into the surface of the clay occur frequently. These are often in the mouth area of anthropomorphic figures, but it is unclear whether they are "mouths" or merely the obliteration of mouths.

Line Forms. The lines themselves fall into two basic elements: straight lines and arcs. Both of these basic forms are extremely common in the cave, even as isolated elements. More common, of course, are combinations of these forms into more complex geometric figures. Straight lines may be linked to form large zig-zag units. Similarly, arcs (that is to say, curved lines) can be combined into various compound elements which I have termed "squiggles" ("squiggle . . . a wriggly twist or curve" [OED]). There are several types of squiggles, depending upon whether the arcs are joined smoothly or with hard angles. In many instances, the squiggles overlap to form closed figures (see Plate VI).

In this report a simple descriptive code is used to indicate some of the features and combinations of forms present in the Mud Glyph Cave art. This is not intended to substitute for a true morphological and structural analysis of the material but is employed for its utility and relative simplicity. The following symbols are employed:

[. . .] Brackets enclose a single motif or analytical unit
    // A slash is used as a delimiter between features

// Double slashes indicate separation of two design units within a
   larger unit
± Plus or minus indicates the state of a feature.

The nesting of smaller and more basic units in larger motifs is indicated
more clearly in the nesting of labeled boxes in Figure 3. These are the
same weakly formal descriptions as used in the text but are graphically
ordered. Note the way in which more complex forms, sometimes quite
different in end appearance, have similar compositions and how a simple change in one feature can alter morphology considerably.

Most of the forms in Mud Glyph Cave can be essentially reduced to
a single element, line (coded with the feature [+line]), which varies ac-
cording to which features it displays. Non-line elements such as "pokes"
or "smears" [−line] do occur, but the bulk of the art is line drawing.
The first feature of line is that of [±straight]. A straight line has the
feature [+straight], while a curved line or arc has the feature [− straight].
A line can be joined with other lines in either a smooth or abrupt fashion
to give the features [+connect /±smooth] (see Figure 3). In the execution of the line [±connect] would relate to whether the incising tool was
lifted or held on the surface in the transition from one form to another,
although the features relate to the morphology and structure of the element, not to the actual procedures of execution as such. Another feature
of line form is whether the form is open or closed [±open]. A form that
is [+straight /−open / . . .] would be a polygon such as a square. A
[+line/−straight /−open / . . .] figure would be a circle-like form. A
complete description of the lines and their combinations in the Mud
Glyph Cave would have to extend this simple example of feature analysis
along the lines shown for a few of the forms in Figure 3, but even this
example may help to illustrate how a more formal understanding of morphological features leads to a more powerful and useful "typology" of
forms.

Of course these forms do not occur in isolation from one another.
These lines are put together into much more complex representations.
Representations that can be identified clearly in areas of less overdrawing are often naturalistic representations. There are no definitely non-
representational figures, although many parts of the cave wall are such
a palimpsest of overdrawing that it is often extremely difficult to identify what the subject of the representation is. Nonetheless, I think it

a.

b.

c.

PLATE VI. a) squiggles (43R), b)
tail (?) squiggles (34L), c)
parallel squiggles (17R).

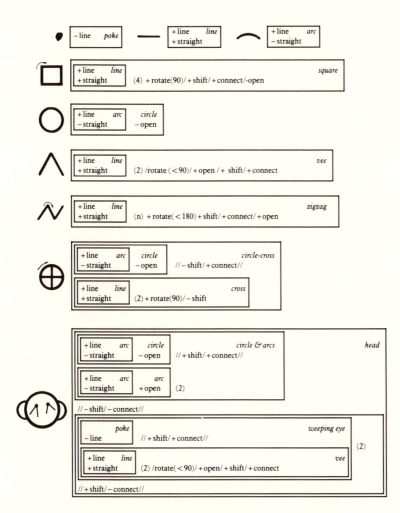

FIGURE 3.1: FEATURES, MOTIFS, AND THEMES.

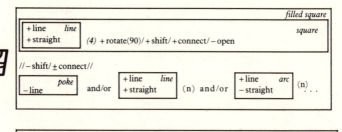

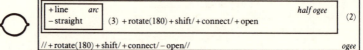

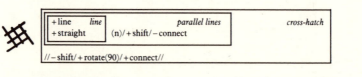

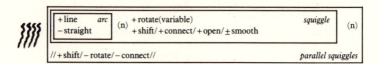

FIGURE 3.2: FEATURES, MOTIFS, AND THEMES.

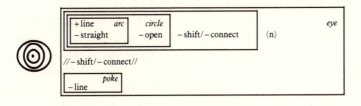

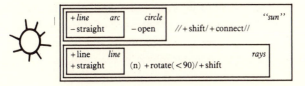

FIGURE 3.3: FEATURES, MOTIFS, AND THEMES.

would be a mistake to simply assume that any of the lines are just random markings on the cave walls. This is especially so insofar as it is clear in a few places that large-scale lines and arcs do form parts of very large figures extending for upwards of two or three meters along the wall of the cave. In addition there are clear instances of use of the natural shapes of the clay surface as parts of designs. Identification of such figures is difficult, especially in photographs which can take in only a section of the cave wall at a time. Angle of lighting and character of the light are also important in understanding the appearance of the cave. The difference between seeing the cave in carbide light and by torch light is great, and the flash needed for photographs is an even harsher and higher contrast light.

Lines and other elements are combined in a number of different ways. One of the most common is to superimpose one set of lines upon another [−shift/+rotate(n) /+connect]. Without [+rotate], the lines would merely be superimposed. The angle of the rotation is indicated by the number in parentheses. Depending upon the angle, this produces various forms of cross-hatching and complex free-form combinations. Lines are also commonly drawn in sets of parallel units [+shift /−rotate /−connect]. In some cases, this was done by actually holding a bundle of three or four canes (or other tools) and making a single stroke. In other cases, however, there is evidence that the parallel lines were carefully drawn in, one at a time. Although the general impression of the cave art is that it was done hurriedly, circumstances such as these suggest that, on occasion, considerable care was taken.

Motifs and Themes

Motifs are consistent combinations of lines and other elements that recur in making up the larger representations. There is no definite division between motif and "theme" except for scale—both are levels of combinations of smaller elements into larger, presumably "meaningful" designs. Both motifs and themes are arbitrary "types" of design that are most useful at a preliminary level of analysis. A full analysis of motifs has to provide the structural and morphological proof for the typology. What follows, however, is only preliminary. It is presented because of its potential utility in leading to a more complete future analysis.

There are a number of apparent themes represented in the cave, but by far the most common combinations of motifs are those that can be generally classed as "anthropomorphic" and "herpetomorphic"—human-like and snake-like figures (Plate VII). Properly speaking, the distinction between these classes makes sense only after the motifs of which they are composed are presented, but since most of the motifs are apparent body parts of the larger figures, they are also discussed in each section in terms of their association with one another in thematic representations. First, however, I will discuss the basic combinations of line from which larger motifs and themes are built.

Basic Motifs. The elements of lines [+line /+straight], arc [+line /−straight], squiggle [+line /−straight /+rotate /+shift /+connect /+open /±smooth / . . .] (see Figure 3.2), and so on were often used alone to form parts of larger units. There are also many cases in which consistent combinations of elements are used as a unit in their own right. Many elements are simply multiplied or duplicated to form motifs. Examples of this include parallel lines/squiggles and concentric circles. In other cases, elements are combined with rotation to form crosses, crosshatch, or other kinds of overlapping motifs. The most basic kinds of design elements—created through various kinds of joining of elements, rotation, shifting, and so on—are relatively few; yet a broad range of more complex motifs and themes were created using these basic forms. The simplest motifs are those that are created through direct use of the elements. Above that level, geometric figures are created such as squares [+line(4) /+straight /rotate(90) /+shift /+connect /−open], circles [+line /−straight /−open], and zigzags [+line(n) /+straight /+open / . . .]. More complex figures were created by combining larger units: a circle-cross is a circle and two crossed lines (see Figure 3.1 for an illustration). Each of the various motifs that occur in making up larger themes can be described in this way as combinations of smaller line units.

Each of these motifs is given a name that is intended to have mnemonic value. While this makes the discussion far easier to follow than assigning the motif a number or code letter, it places a burden on readers of continually reminding themselves that these terms do not have a necessary relationship to the "meanings" these drawings had for the prehistoric artisan. Thus the term "face" is used to describe certain sets of lines and arcs, but the referent may not have been conceived in that fashion by the artist who may have seen the motif only as a part of larger thematic representation or as a symbol of some deity or spirit. While

a.

b.

Plate VII a) anthropomorphic glyph (70L), b) herpetomorphic glyph (87R).

it may not be completely impossible to relate the symbols of the cave to those of historic Native Americans, such an effort has to be undertaken with scientific rigor, not with an anecdotal "this must have been . . . " approach. At the risk of seeming to nag the reader on this topic, I have to urge the greatest caution in assigning "meanings" to these motifs and designs. If the "serpent" representations in the cave are to be identified as the Uktena or the "winged humans" as the Tlanuwa of the Cherokee, then it must be shown that the cave materials were actually done by a Cherokee, that the features in the cave show a consistent combination of forms that fit that meaning best, and that changes in meaning did not occur in the time between the cave drawing and the writing down of the myths. These are high standards, but there is little point to calling the "serpent" an Uktena if it really had different semantic and cultural "meaning" than the historic Cherokee monster. It is for these reasons that I have frequently used the terms anthropomorphic (human-like) and herpetomorphic (serpent-like) to describe two major classes of representation. As awkward as the terms are, they serve to remind us that the figures were probably not specifically human or snake to the person or persons who drew them.

Anthropomorphic Motifs. The class of anthropomorphic motifs is largely made up of various combinations of body parts used as eyes, mouths, limbs, and so on. Each of these motifs, however, was used in many different circumstances, perhaps suggesting multiple referents. The most obvious anthropomorphic motif is the "head." These are the easiest motifs in the cave to recognize, many being essentially like a "happy face"—a circle with enclosed eyes and other elements (see Figure 3.1 and Plate VIIIa). A second form of head is the "profile face" (Figure 3.2, Plate VIIIb), often drawn quite freely with little effort to distinguish such profile aspects as nose and chin. The effect of these profiles is very much like the drawings in children's books of the profile view of the "Man in the Moon." Both full and profile heads were done with straight lines [+straight] instead of curved lines [−straight], yielding square faces (Plate VIIIc). Many "squares" (see below and Figure 3.2), may have been intended to represent faces.

As can be seen in Plate VIIIc (and Figure 3.1), the faces often have appended arcs that might represent ears but could be earspools or some kind of hair arrangement or ornament. In any case they are simply arcs

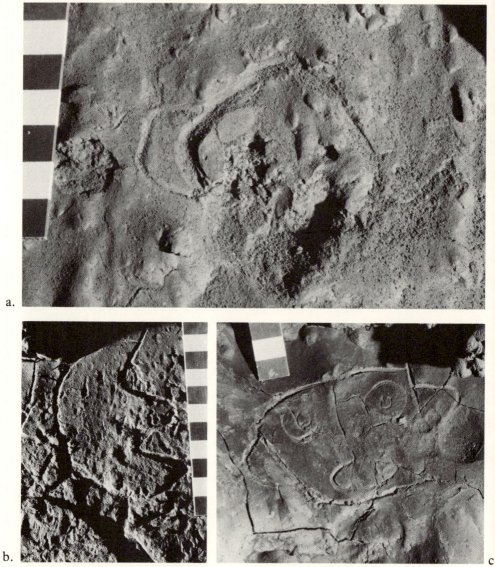

PLATE VIII. Head motifs: a) round (92L), b) profile (73L), c) square (56R).

attached to the sides of the head. Within the head circle there are normally "eye" motifs. Common markings for the eye include pokes [–line], diamonds (like squares with [+rotate($\leq$90)], see Figure 3.1), or almond-shaped units (somewhat like a simplified ogee, see Figure 3.2). Circles also occur in this location on the field. All kinds of eye motifs may have appended inverted "vee" patterns (see Figure 3.1 and Plate IXa) which form the Eastern Woodlands "weeping eye" pattern of great areal spread and duration. More complex weeping eye patterns can have compound curves like ogees (Plate IXb). The eye location may simply be pokes or jabs, as noted; but some of these, at least, may be later defacement of eye variants discussed above. In some cases, natural protrusions and "blobs" of clay on the wall were used as eyes (or other elements).

Many of the full-face motifs lack any clear "mouth," but some round heads do have scratched out areas or a horizontal line below the eyes. A few full-face mouths are ovals with apparent teeth represented by vertical lines. This area of the head is usually obliterated by scratching or by blows to the clay. In those cases where the junction of the head and bodies has not been obliterated, the "neck" is often two straight parallel lines. In some cases, straight lines, sometimes cross-hatched, might be "necklaces" of the sort seen on shell gorgets from the region.

The main portion of the body trunk on the anthropomorphic figures is also commonly disfigured. The trunk is generally reduced in scale relative to the head and appears to have been executed quickly and without great care (Plate X). The body is often delimited by two or more parallel, vertical lines with cross-hatched or squiggle lines inside the body area. Because of overdrawing, it is often difficult to define body limits clearly. Upper limbs are more easily distinguished. In nearly all cases, the arms are outstretched, usually with the lower arms raised up. In most cases, also, there are lines or squiggles running down from the arm. These appear to be stylized representations of "wings" such as those seen on shell gorgets from the region. In some cases, the lines look very similar in form to the "feathers" shown on the gorgets. Another similarity to the so-called "eagle-warrior" gorgets is the termination of limbs in "claw"-like elements rather than in hands and feet. In the cave, however, all the claw-like units are vee patterns or radiating lines, rather than the carefully executed bird claws of the gorgets (see Hightower style in "Relationship to Other Styles and Themes," below). Feet, being in areas that are often drawn over repeatedly, are more difficult to distinguish but

a.

b.

PLATE IX. "Weeping eyes" a) inverted vee (62L), b) elaborate weeping eye (15R).

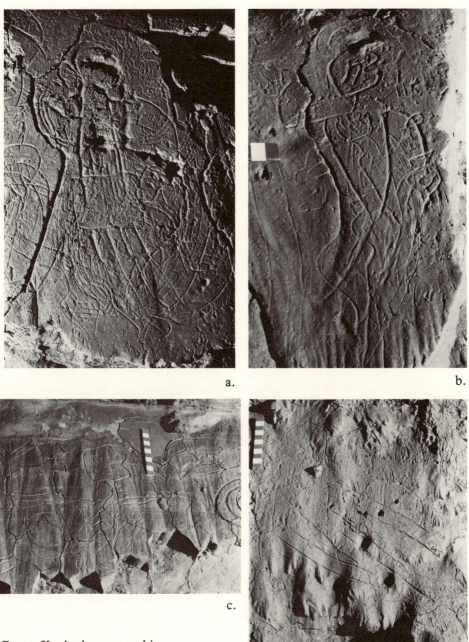

a.

b.

c.

d.

PLATE X. Anthropomorphic
figures: a) (69R), b) (66R), c)
(87L), d) (60L).

appear to be similar to hands in most respects. It is fair to say that very little effort was expended on making clear feet. The vee pattern used in the hands and feet is essentially identical to the vee pattern used in weeping eye motifs. Lower limbs are most often merely extensions of lines that also delimit the main portion of the body. In a few cases there are well-defined legs, and they are characteristically in the bent knee position found on shell gorgets throughout the Southeast. There are a few cases of what appear to be lines and pokes which might represent genitals at the lower portion of the trunk. The use of arc lines to draw these items makes it surprisingly difficult to tell whether they were supposed to represent penises or vulvas! It is clear that if so basic an attribute of these figures is ambiguous, interpretations of the motifs have to be carried out with extreme caution.

In general, the motifs employed in depicting anthropomorphic figures are similar to those found in copper plates or shell gorgets from the upper Tennessee Valley, but the style of representation is most definitely different. As suggested, this may indicate the isolation of the cave in social or geographic terms; or it may simply mean that the styles used in the cave (and perhaps, even probably, on clay plastered walls) were different than those used on more costly materials. Altogether, the anthropomorphic figures are the most common representations in the cave. In at least one case a human figure with outstretched arms appears to be attached to a large bird-like tail (near the 58 meter stake on the left side, indicated by "58L"), like those seen on engraved pottery from Alabama and on the gorgets from eastern Tennessee and northern Georgia. Of course the weeping eye is also seen by many as being a bird-related motif, further supporting the identification of the figure as the same personage or "monster" as on the gorgets.

Herpetomorphic Motifs. Many of the same motifs and elements also appear in the so-called "serpent" representations in the cave. "Eyes," for example, are often represented by circles or diamonds. Similarly, squiggle lines are used in ways that may suggest wings on the serpents as they do more clearly on the human representations. One major difference from the anthropomorphic drawings is that the serpent-like (or "herpetomorphic") representations are often very large, covering meters of cave walls. This, combined with obliteration and reuse of areas, makes it very difficult to identify the entire representation. Many apparently isolated eyes or other elements may actually be part of huge, borderless herpetomorphic figures. There is every reason to believe that natural boundaries of clay areas and of the cave wall were incorporated into these large drawings in the same way as smaller natural protrusions were employed in anthropomorphic drawings.

The eye represented on these forms is most commonly a series of concentric circles (Figure 3.3 [circle //−shift /−connect//(n) //−shift /−connect //poke]). The concentric eye motif may foreshadow concentric eye patterns on "rattlesnake" gorgets of the sixteenth century and later on, but this form of eye is widely documented from many areas of the Southeast in the thirteenth century and cannot be used to argue either for a sixteenth century date for some of the cave drawings or for any necessary connection to the later shell gorget styles of the same area. Patterns such as diamonds, weeping eyes, and single circles are also used in Mud Glyph Cave in long combinations of lines and squiggles and probably represent eye units here as they do in the anthropomorphic drawings. Unlike the later prehistoric and early historic rattlesnake gorgets, the herpetomorphic figures in Mud Glyph Cave do not have chevron elements in tail or body areas and do not have distinct mouth areas like those of the gorgets.

The bodies of the more-or-less distinct herpetomorphic figures are made up of both repeated vertical lines, often roughly parallel to one another, and parallel line squiggles (see Figure 3.2; Plate XI). Crossedover, single-line squiggles appear to represent "rattles" at the ends of these figures, and there are radiating arcs from the eye area which may indicate some kind of "crest" or even "wings" such as seen on engraved pottery herpetomorphic figures in the Moundville region. In a few cases, horizontal lines delimit the body of the figure, but it is more common to have no clear limits other than those imposed by the natural forms of the wall.

The motifs used to make up the serpent representations in the cave are similar to many apparently isolated elements in the cave, so it is possible that many larger herpetomorphic drawings are unrecognized or obscured. At least two serpent representations are also apparent "bird" heads. In these cases it is possible that one sort of figure was later modified to convert to another purpose, but it is not impossible that the ambiguity was intentional. This kind of multiple "reading" symbolism is, of course, a common occurrence around the world, especially where the

a.

b.

b.

PLATE XI. Herpetomorphic figures: a) (78L), b) (57R).

symbols are of ideological significance. In historic times in the Southeast, combination of serpent, human, and bird attributes was common (as, for example, among the Creek and Cherokee cultures). Such ambiguity could have been the result of interaction of the artist with the drawing-in-progress, or it could have been the result of careful planning. The casual or cursive character of much of the drawing perhaps indicated that the former alternative is to be preferred. Moreover, we have no clear idea of how common the use of these elements and themes may have been. The location in the cave seems to suggest that these themes were "secret," but we really cannot know whether the themes were widely used and appeared on every house wall. The crude appearance of the drawings at first glance is belied by their sophistication of combination and representation on closer examination. Of all the themes, the large scale and character of the herpetomorphic drawings make them the most difficult to study and most susceptible to the "Martian" problem referred to above.

"Geometric" Motifs and Themes. As can be seen, all of the motifs in Mud Glyph Cave are in some sense "geometric," but some of these occur in apparent isolation of other units and cannot be linked easily to some larger drawing. Perhaps the commonest ones of these motifs are the various circle, inverted vee, and diamond (or, properly, rhomboid) patterns found throughout the cave. Identification of the inverted vees as "weeping eyes" seems unjustified (especially insofar as these same lines can indicate "claws" as well). Circles and diamonds, as noted, are common eye patterns as well but may have had different referents as stand-alone motifs. Another common class is that of whole and partial ogee patterns (Figure 3.2; Plate XII).

Ogees are basically a rotated and joined compound curve. In Mud Glyph Cave, these are commonly drawn with multiple, parallel lines as seen in Plate XII. This motif is common in other media from various parts of the Southeast, either on its own or as an eye, anus, or vulva. It does not clearly seem to represent these body parts in Mud Glyph Cave glyphs.

Simple arcs of parallel lines are common as well, sometimes in association with apparent serpents. Single line designs include the circle-cross (Figure 3.1). This motif is very wide-spread and ancient throughout the New World (not to mention the Old World). Interestingly, it is not com-

PLATE XII. Ogee (37L).

mon in the cave, although it probably occurs more often than any other motif in the art of the Eastern Woodlands known to us from other sources. Circles also occur together with radiating lines to make a "sun" representation (Figure 3.3). Radiation lines also occur without a circle. There is no clear indication of a cross within a "square" unit, but otherwise squares and rectangles are treated essentially the same way as circles are. Both circles and rectangles, for example, may have internal filling of lines and arc elements (Figure 3.2; Plate XIII). Very large, "Easter egg"-like units appear to be similar motifs made with arcs rather than straight lines. Other star-like geometric figures occur which have some similarities to engraved shell materials from the South Atlantic Coast region (see discussion below).

"Southern Cult" Motifs and Themes. Although many have tended to treat all elaborate artistic materials of the Late Prehistoric Southeast as being "Southern Cult," it seems more useful to recognize that there

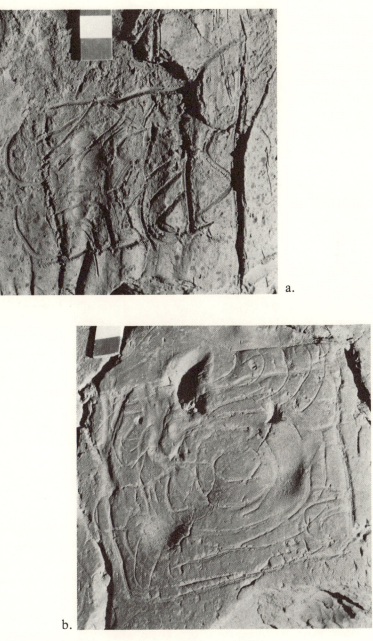

a.

b.

PLATE XIII. Filled rectangles: a) (36L), b) (68R).

were several more narrowly defined horizons within this time period (Muller in press, volume 13 of the *Handbook of North American Indians*). Many of the so-called "markers," such as the weeping eye motif, that have sometimes been used to define supposed "Southern Cult" association are so widespread in their temporal and spatial distributions as to be useless for that purpose. There is, however, a core of motifs that is limited to the later thirteenth-century horizon that everyone agrees is "Cult." The most distinctive of these horizon markers is the bi-lobed arrow. Many suggestions have been made for the referent of this motif, ranging from a symbolic atlatl (spearthrower) to male genitalia. Whatever it meant, there is one fairly clear bi-lobed arrow in Mud Glyph Cave at 41L (see Plate XIV). There is another possible bi-lobed arrow shown on the back of a head at 55R. At 41L, the bi-lobed arrow part of the drawing is the lowest of a complex combination of a human figure, and possibly a mace as well. Mace-like motifs also occur in several locations along the left wall. The mace motif is also a fairly good marker for the mid- to late-thirteenth century Southern Cult proper.

Other Southern Cult items such as a striped pole motif have not been identified in Mud Glyph Cave art. The most important of these is the "apron"-like motif that commonly occurs with human figures everywhere in the Southeast, including the immediate region of Mud Glyph Cave. Its absence here does parallel its rarity in regionally made gorgets, however. All in all, the evidence discussed here, and in the sections below on relationships to other styles, suggests that the radiocarbon dates are basically correct for the dating of the main episodes of glyph drawing in Mud Glyph Cave.

Animal Representations. The small owl at 63L is one of the most charming, and most recognizable, drawings in the cave (Plate XV). Nonetheless, it is untypical in many respects. Its small size and the fineness of the lines used in drawing it are rare in the cave. In addition, there are relatively few animal representations in Mud Glyph Cave except for the serpent motifs already discussed. There is a small drawing of a probable opossum at 93L. Probable bird elements, as discussed, are commonly associated with both the anthropomorphic and the herpetomorphic figures in the cave. There is an isolated "hawk" head at 54R (Plate XVI). At 30L, one glyph looks very much like the woodpecker heads on shell gorgets, and it is likely that this is either a

PLATE XIV. The bi-lobed arrow/mace/human
figure (41L).

converted human figure as well, or that its reference was purposely ambiguous (Plate XVII). The glyph at 33L has some woodpecker-like
features but appears to have been mainly a serpent representation.

As already noted, a large percentage of the squiggles and lines in the
cave may well indicate bird or other animal characteristics such as
feathers, wings, or even skins or horns. There are also two double circle
glyphs that may once have been spider drawings like those seen on shell

PLATE XV. Owl at 63L.

gorgets from the central Mississippi Valley, but these were so badly
damaged that it is not possible to suggest such identifications with any
degree of certainty. Figures formerly identified as "turtles" (e.g. the
elements at 56R) seem more likely to be human heads of the full-face
variety, although it is certainly possible that there was intentional ambiguity here, as in other cases. Even taken altogether, however, identifiable non-human/non-serpent animal representations in the cave are
rare.

"MEANING"
One thing is ironically clear: many of the drawings seem to have been
purposefully ambiguous in reference. Numerous drawings were either
constructed to suggest more than one "meaning" or were later modified
to represent a different figure. Given the strong likelihood that there
is no easy, one-to-one mapping of referent to drawing, the dangers of
facile interpretation are obvious. Not only is the intention of the artist
obscure, but the ease with which the analyst can create chimera (literally and figuratively) out of his own perceptions (the "Martian" problem,
again) shows the need to establish the identifications of these

PLATE XVI. Hawk head at 54R.

glyphs only after careful study and comparison with other, contemporary materials. Comparison to historic mythology is not useless nor foolish but is certainly premature until we have some basis for identifying classes of referents and can describe their associations and features. For example, there is no doubt that many societies in the Southeast had winged/ horned/spotted serpents and horned/winged human-like figures in their mythology, but it would be unwise to identify these historic mythological figures too closely with specific cave motifs and themes. It seems particularly unfortunate to use Cherokee names for these themes, since however fully the Cherokees may have integrated these elements into their myths, linguistic evidence suggests that they were relative newcomers to the area. The symbolism of the Mud Glyph Cave has to be interpreted and studied in its own terms, especially since the style of representation is unique to this site. The materials in the cave are not stylistically similar to historic nor protohistoric Cherokee art, in any case; and they share only general Eastern similarities in themes. It scarcely

PLATE XVII. Woodpecker/human at 30L.

seems better motivated to use historic Creek or Yuchi interpretations, either.

Even though it seems dangerous to identify the Mud Glyph Cave figures as Uktena and Tlanuwa, the association of rattlesnake-like figures with the underworld in much of the Southeast is clear (see Hudson 1976 for an excellent summary of Southeastern "monsters"). In addition, themes like the "Piasa" figures (Phillips and Brown 1975–82:140) are very common in both historic and prehistoric art from the area. At the very least, it seem profitable to hypothesize that dualist representation of opposing, perhaps Upper and Lower World, figures were the subject of much of the art in Mud Glyph Cave.

RELATIONSHIP TO OTHER
STYLES AND THEMES

EASTERN TENNESSEE VALLEY STYLES

Hightower Style (a.k.a. "Mound C style"). This is the "typical" style of Southern Cult times (Muller 1966b; in press; Phillips and Brown 1975–82:182) in eastern Tennessee, and it is no surprise that the most specific similarities of the Mud Glyph Cave art are to the themes on these gorgets. This is the style that is best known as that on the gorgets recovered from the mortuary complex in Mound C at the Etowah site in northern Georgia, but the style also is well represented from many sites south of Knoxville and north of Chattanooga in the eastern Tennessee Valley. The most common shell gorget figure is that of the "winged" anthropomorphic figure (Plate XVIII).

The representation combines wing elements; antlers; bird-like tails; and clay hands, feet, or both. The bi-lobed arrow is present, usually with circular lobes and often with a flame-like top. Although the figure often has a "dancing" position, the known figures are all upright, rather than sideways as in specimens from further west (see Eddyville style). The most common form is with outstretched arms in exactly the same positions found in the Mud Glyph Cave drawings. The arms of the gorget figures are usually linked to or associated with wing elements, as is also generally the case in Mud Glyph Cave (although most clearly represented in 60L, 87L, and 66R). The termination of the limbs in the cave is usually in a "vee" pattern of some sort and this seems to correspond to the claws on the Hightower figures (60L and elsewhere). The generally upright position of the human figures in the cave is also a characteristic of the Hightower gorgets. The profile faces in Mud Glyph Cave are also very similar to those of the Hightower anthropomorphic theme, without being very similar stylistically. The circular or arc "bun" pattern on many of the Mud Glyph Cave figures has no precise analogue in the Hightower gorgets, unless the arcs in the cave represent earspools which are universal in the gorgets. The antlers of the Hightower gorget figures may be what was intended by some of the lines radiating from the heads of anthropomorphic heads in the cave, but the treatment in these is much

PLATE XVIII. "Hightower" gorget from Hixon site, Hamilton County, Tennessee (University of Tennessee 617/1Ha3).

more similar to that seen on the later Williams Island gorgets of the main Tennessee Valley and from northern Alabama.

As for other similarities, the tail of the human figure at 58L (Plate XIX) is as much like the tail on the Hightower gorgets (see Plate XVIII) as it is like Moundville engraved pottery motifs. Indeed the Hightower and the Moundville "bird" tails are very similar to one another, a fact appreciated only as a result of this study. It is also likely that the Cox Mound theme (four woodpecker heads on a parallel line "square") is more closely related to the Hightower style than thought earlier. Just as the northern Alabama area lies between the geographical areas of the Eddyville and Hightower styles, so does it seem to form a stylistic transition between the two styles. It should be cautioned that there are relatively few gorgets in the northern Alabama and Moundville localities, however, from the peak Southern Cult period.

Most of the "human" representations in Mud Glyph Cave do not

PLATE XIX. Anthropomorphic figure and bird tail (58L).

have clear necklaces or bracelets like those seen on most gorgets, but the figure at 66R has a necklace, bracelets, and armbands (Plate Xb). The Mud Glyph Cave depiction of these elements is by simple lines which does have a few parallels in other media but lacks the brickwork squares seen commonly on Eddyville gorgets or the three parallel-line units or circular "beads" seen on Hightower gorgets. The Mud Glyph Cave human faces do not have a clear forelock element associated. Another feature of the Hightower human gorgets that is lacking on the Mud Glyph Cave figures is the absence of circle-and-dot and "web" patterns in the crotch. The Mud Glyph Cave figures almost invariably have simply a "poke" or lines in the genital area. In at least one case, the lines appear to be a representation of a penis, a feature totally absent from any gorget representations.

In terms of larger scale themes, the spider theme is rare in the Hightower style (the one example is from Etowah [R.S. Peabody Museum 61389])—unlike the western Eddyville style (see below)—and may be absent from Mud Glyph Cave. As indicated, one possible spider in Mud Glyph Cave occurs in a three-circle unit at 58L, but this is by no means clearly a spider. Although spiders are rare in the mid-thirteenth century Hightower style in the eastern Tennessee Valley, at least five spider gorgets are known from a later style that appears to have developed out of the Hightower style. The most striking difference in themes between Mud Glyph Cave and the Hightower style, however, is to be found in the complete absence from the cave of any representation of turkey cocks—an extremely common theme in Hightower and in its descendant styles. Nor are owls represented in any gorget style. Of course Mud Glyph Cave, as a cave, may have had fearful associations of monsters, and it is not to be expected that the use of a cave was related to the same portions of the belief system as were artifacts associated with other places and activities.

Bennett Place. "Square Cross" gorgets show the cross in a circle which is bounded by a square with projecting lines. The theme is named after the Bennett Place find of C.B. Moore (1915:347, Figure 74; also see Kneberg 1959:Figure 1) [Museum of American Indian 17/1306]. This form may have begun relatively early in Mississippian times, but it also occurs in later Mississippian contexts. For example, it occurs in the grave lot at the famous Castalian Springs site in Tennessee (where the square encloses a woodpecker instead of a cross in the center, Museum of the American Indian 15/8055). There are other specimens that seem to be late, so the motif is unlikely to be a good horizon marker. No direct example of the theme is represented in the cave, but the arcs and circles in squares are somewhat similar.

Shell Masks. There are a few similarities, most notably some of the "skull" (70L) figures look a little like some of the (probably early) shell masks (specimens from Crittenden County, Arkansas [Gilcrease Museum] and the Field Museum 51541-9, supposedly from the Spiro site). Note that the masks have hole eyes and sometimes mouths, as do some of the full face, circle figures in the cave. At least one cave figure appears to have the fringed border found on most shell masks. A double line border is found in 70L, but that may merely be a parallel line figure. The best and most shell-like figure is at 62L (62L#2, Plate IXa), which is very much like shell masks. As I have noted elsewhere, the shell masks were most common after A.D. 1450 or so, but there is much to establish that the motif and the objects themselves were present back to the A.D. 1250 period. Thus, although the appearance of mask-like figures might suggest the lateness of the cave glyphs, it is not inconsistent with the radiocarbon dates in the mid-thirteenth century.

South Atlantic. There are similarities between the geometric patterns at 54L and 69R (Plate XX) and some questionable materials from western Virginia in the Saltville area (in private collections). Other glyphs are similar to the designs in drilled pits from a "South Atlantic" style at the Irene site and elsewhere (Caldwell, McCann, and Hulse 1941:Plate XIX; Muller 1966b) and nicely executed specimens from Florida (Rouse 1951:Plate 6). The similarities are not overwhelming, and most of the South Atlantic materials seem to be later than the probable date of Mud Glyph Cave.

Williams Island and Hixon. In the eastern Tennessee Valley, there seems to be a clear evolution of art styles between the mid-thirteenth century and the mid-fifteenth century. I have elsewhere given the names Williams Island and Hixon to the themes that derive from the earlier, Southern Cult horizon styles of the region. Williams Island refers to the anthropomorphic theme, and the Hixon style themes include both the turkey cock and the spider. It is probable that the Williams Island and Hixon themes are related, but their direct stylistic connections have not been established. As noted, the spider and turkey cock themes are absent from Mud Glyph Cave, so there are no direct parallels between the Hixon style group and the cave materials. In the case of the anthropomorphic theme characteristic of the Williams Island gorgets, there are a number of similarities. The Williams Island gorgets were made later than the more naturalistic Hightower gorgets, and various elements present as distinct motifs in the earlier treatment of the theme are reduced to mere parallel line units in the same position on the gorget field. Because of the overlapping and "baroque" character of the treatment, these are often called "spaghetti" figures (Phillips and Brown 1975–82:197). This baroque version of the anthropomorphic theme appears to continue to be made in northern Alabama up to protohistoric times; but, in the eastern Tennessee Valley north of Chattanooga, the use of this theme appears to stop with the spread of the various "rattlesnake" gorget themes (q.v.). Both the Mud Glyph Cave and the Williams Island representations are reductions of elements executed more carefully and

PLATE XX. Geometric pattern at 69R.

naturalistically in the Hightower style, but most of the similarity ends there. The specific treatments of head, arms and legs, eyes, mouth, antlers, and the like are entirely different, even though the theme is likely to be the same. This difference further supports the mid- to later thirteenth-century dating of most of the Mud Glyph Cave art.

Nashville Style. These are the "scalloped triskele" gorgets of Kneberg (1959). They are most common in the area surrounding Nashville, but they occur from North Dakota to Georgia. They are very common in mid-fifteenth century contexts in eastern Tennessee. Associations in various parts of the Southeast appear to place their peak of popularity in the years around A.D. 1450, but the theme itself was present as early as the late thirteenth century in other media such as negative painted pottery (Nashville Negative; Nashville Negative, variety Angel, see Black 1967:474 and elsewhere) except that the central figure was more often a cross than a triskele. I believe that the scalloped triskele form is a relatively good horizon marker for the mid-fifteenth century over much of the Southeast, and its complete absence from Mud Glyph Cave further supports the dating of the cave as mid-thirteenth century.

The "Rattlesnake" Gorget Styles. The Lick Creek, Citico and Saltville styles are the most formally analyzed of the shell gorget styles in the Southeast (Muller 1966a). These three styles, unlike the others discussed here, are based on detailed analyses of both theme and form. The most obvious similarity of these objects to the Mud Glyph Cave art is the use of the herpetomorphic theme. Having said this, the fact is that there is really essentially no similarity between the Mud Glyph Cave and the various "rattlesnake" representations on the gorgets. The Mud Glyph Cave serpents are much more like those on Moundville engraved pottery than they are like the shell gorgets from the eastern Tennessee region. Most of the rattlesnake gorgets have a strong "cross" orientation, and both chevrons and cross-hatching are crucial elements—chevrons are completely missing from the cave and cross-hatching is treated differently. Concentric circle eyes are a feature of both the gorgets and the Mud Glyph Cave serpents, but there is virtually no other similarity between the two that is not explainable by reference to the appearance of real snakes.

OTHER REGIONAL STYLES

Moundville. Although there are many apparent similarities between Moundville ceramic engraving and the Mud Glyph Cave, there are essentially no similarities between the few human figure Moundville gorgets and the cave glyphs (Phillips and Brown 1975–82:195ff.). The woodpecker at 30L has points of similarity to Cox Mound, "woodpecker" theme gorgets (C.B. Moore collections, Museum of American Indian 17/2271, 17/929). Cox Mound theme gorgets are known from eastern Tennessee and may be exchange items from northern

Alabama (one specimen, United States National Museum 388037, is attributed to a site relatively near Mud Glyph Cave—the Tellico Mound). The tail which is apparently attached to the human figure at 58L (Plate XIX) is very like some of the tails on Moundville engraved pottery, but there is no corresponding motif known in Moundville engraved shell.

Eddyville. This is the western Tennessee-Cumberland-Mississippi style. It is known from the Nashville area west to northeast Arkansas. It is essentially the same style as the "Braden A" of Phillips and Brown (1975–82). Stylistic correspondences to Mud Glyph Cave are only general, but there are a number of thematic similarities. The parallel-line borders of most Eddyville style gorgets might be similar to the use of three and four parallel line units in the cave, but this is stretching the data a bit. Other similarities, in no particular order, include portrayal of human "dancing" figures in a sideways position. This is found in the "jogger" in the Mud Glyph Cave 69R (Plate Xa) typical of Eddyville gorgets from Arkansas (2 figures in a sideways "court card," McGimsey 1964:Figure 3), Illinois (see Phillips and Brown 1975–82:174ff.), and other parts of the Tennessee-Ohio-Mississippi confluence region. The Castalian Springs gorget from north central Tennessee has a number of thematic similarities to the cave drawings. The arms are partly up, though not as much as on the "jogger." The bi-lobed arrow on the head could be paralleled in 55R. The mace on that gorget is similar to that at 41L. The "trophy head" in the right hand of the Castalian Springs anthropomorphic figure may have some similarity to some circle faces that seem to be associated with winged figures in the cave (e.g. some of double face figures at 63L, 45L, 58L and note the crude head at the waist of 69R, Plate Xa).

A difference between the Eddyville materials and the cave art is that there are no clear examples of spider representations in the Mud Glyph Cave art. The spider theme, however, is only common in the Illinois and eastern Missouri portion of the Eddyville style region. It should be noted that many of the similarities between the Eddyville materials and the Mud Glyph Cave are also points of similarity between the Eddyville style and Hightower style of the eastern Tennessee Valley (q.v.). One important element missing in the Mud Glyph Cave drawings is the "apron" which is so common on Eddyville "human" gorgets. As noted, this is not surprising, as the apron is also rare in the eastern region shell gorgets. There is one possible, but doubtful, "apron" at 58L in the Mud Glyph Cave.

The Spiro Styles. The Spiro materials have been presented in a magnificent publication from the Peabody Museum at Harvard (Phillips and Brown 1975-82) comprising six volumes—five of them large-scale reproductions of rubbings of the shell surface. As I have noted in a review of these volumes (Muller 1984), they provide us with a corpus of material that is more varied and extensive than for any other area in the Southeast. In fact, in all of the Southeast, only the Mud Glyph Cave offers us a similar "sealed" deposit with comparable extent of representation.

At least one of the shell cup and gorget styles at the Spiro site is essentially the same as the Eddyville style discussed above (Phillips and Brown 1975-82:179). For reasons which I have hinted at elsewhere (Muller 1984), I am reluctant to use the name "Braden A" as a style name for gorgets in the Lower Tennessee and Mississippi Valleys. Phillips and I, however, are not in disagreement about the essential characteristics of the style in question, and what has been said above about the Eddyville style also applies to this material at Spiro. Other shell cup and gorget engravings at Spiro are much less like the Eddyville material, and their similarities to the Mud Glyph Cave art are less than that found in Braden A or Eddyville.

In the section that follows, the "Glossary of Motifs" for Spiro art listed by Phillips and Brown (1975-82:145-156) for Spiro is compared to the art from Mud Glyph Cave because these two sites provide the largest body of data on the "Southern Cult" horizon. As will become clear, there are relatively few exact correspondences between the two sites. Two kinds of similarity exist: those that are the result of widespread use of certain motifs during the Southern Cult Horizon and those that are present in both areas as a result of exchange relationships between the two areas. Of the "Glossary of Motifs," the following observations may be made:

Agnathous Head—no representation in the Mud Glyph Cave art.
Akron Grid—lines terminating in a line or triangular element. This is noted at a number of locations in the Mud Glyph Cave, but only the individual element of a line ending in the unit is noted and these may simply be "serpent" elements that happen to look like the units of the Akron Grid (36L, 66R, others).

Amphisbaena Tongue—no representation in Mud Glyph Cave.

Antlers—there are many probable antlers in the cave, but all are represented merely as lines coming out of the head. None is drawn as a clear antler comparable to those of Braden A, Hightower, or other shell art styles.

Arrows, Arrow Feathering, Arrow Points, Arrow Snakes, and Arrow Tongues—not present in any of the Mud Glyph Cave art.

Barred Oval—this unit is not clearly represented, but some bisected circles could be a similar motif and the pokes seem to be comparable in location and function.

Barred Rectangle and Bellows-shaped Apron—lacking.

Bi-lobed Arrow—rare, as noted, in the Mud Glyph Cave. The clearest unit of this motif in the Mud Glyph Cave art is of the "bi-lobed plume" sort.

Bird Head, Bitriangular Arrow, Bow, Bowknot, Broken Arrow, Broken Mace, Broken Staff, and Carrot-shaped Appendage—all absent.

Concentric Circles—rare at Spiro, this is one of the most common forms at Mud Glyph Cave. Concentric Semicircles (arcs) are also common at the Mud Glyph Cave but restricted to the probable earlier (Braden A and Craig A) phases at Spiro.

Concentric and Single Bisected Semicircles—as noted above, bisected circles are common as parts of human and other figures in Mud Glyph Cave art. Phillips and Brown (1975–82:149) also include here an illustration of a "shield-like" object that is similar to the squares filled with arcs and circles of the Mud Glyph Cave.

Concentric Radial T-Bat Motif—this is the pattern noted on the crotch of the anthropomorphic figures of the Hightower style. Its associations at Spiro are different, but the form is lacking in the Mud Glyph Cave.

Crossed Bands—not present in Mud Glyph Cave.

Cross in Circle—this is a worldwide artistic representation, and it is present in Mud Glyph Cave as it is in nearly all New World art.

Cross in Petaloid Circle—this is possibly a precursor of the scalloped cross element that is common slightly later in association with the scalloped triskele gorgets. No clear examples are present in Mud Glyph Cave.

Davis Rectangle—a motif mostly associated with Lower Mississippi Valley ceramic styles, but which also occurs in Braden A. This is remotely similar to a reversed curve "key" (or squiggle?) pattern that does occur in Mud Glyph Cave, but no actual Davis rectangle representations are known to me in Eastern Tennessee.

Diamonds, Single and Nested—although nested forms are not present in the cave, the single diamond pattern is a common eye pattern throughout the Southeast, and it occurs on a number of the human and other representations in Mud Glyph Cave.

Dotted Single and Concentric Circles—these are common eye representations which may have other uses. Many of the circle eyes in the Mud Glyph Cave have dotted circle eyes.

Ellipsoidal Eye Surround, Entwined Bands, and Excised Triangles—not present.

Eye—a "dotted diamond," this is another Southeastern universal form. It occurs in the Mud Glyph Cave where it grades into a "two-arc" eye pattern.

Forearm Bones—not present in the Mud Glyph Cave.

Forked Eye Surround—this motif does occur in various forms in the Mud Glyph Cave, but most cases are actually "forked eyes" rather than merely forked "surrounds." The forked eye motif covers the time period from Adena to Historic and cannot be used as a marker for "Southern Cult" time associations. Most forked eyes at Mud Glyph Cave are simply inverted vee patterns, as noted.

Greek Cross—not present.

Hand—no naturalistic hand like those of the gorgets is seen in the cave, but both the "jogger" (69R) and the figure at 66R have crudely drawn hands with stick fingers. The hand itself (or what it is holding?) is sometimes circular (at 66R, two concentric circles-an "eye-in-hand" pattern?). No cross-in-hand is present.

Key-sided Mace—not present.

Looped Square—this is the form associated with the Cox Mound "woodpeckers" in northern Alabama. Although there are similarities in other associated motifs, the looped square itself is not present.

Mace—as noted, this is present in the Mud Glyph Cave, but this is another motif found throughout the Southeast.

Mace and Arrow Feathering and Moundville Circle—not present, but the arcs in squares are very similar to the Moundville Circle, given the general interchangeability of squares and circles in this area.

Nested Rectangles—none of the "squares" in the Mud Glyph Cave are nested.

Ogee—clearly present in the Mud Glyph Cave, but none of the representations there has the oval in the center that is common elsewhere.

Ovals—Mostly missing from the cave, although there are partially oval
figures (e.g. 45L).

Pear-shaped—this form occurs as the shape of the head on a few of the
full-face human figures in Mud Glyph Cave. Where it occurs, it seems
more likely to be a sloppily executed circle than a separate motif,
however.

Raccoon—Although there are a couple of possible cases of raccoon and
raccoon hindquarters in Mud Glyph Cave, none of these are clear
at all.

Of the remaining 12 or so motifs listed by Phillips and Brown
(1975–83:145–156), none is represented in the Mud Glyph Cave, with
the possible exception that some of the full face figures in the cave may
represent skulls, but the appearance is quite different from the profiles
seen in Craig C.

Thus, taken altogether, there are really very few specific similarities
in motifs between the Mud Glyph Cave and the Spiro styles. The styles
are not only different, but the subjects themselves and the attributes
associated with them are rather varied across the Southeast. Given that
the Mud Glyph Cave and Spiro each represent a large and varied collec-
tion of so-called "Cult" materials from roughly the same period (at least
for Braden A), we are justified in being very skeptical of interpretations
that assume or emphasize ideological and religious unity in the Southeast!

DISCUSSION AND CONCLUSIONS

Mud Glyph Cave stands as a classic example of the problems of archaeological interpretation. It provides an extremely "fine-grained" set
of data in that its art probably represents a short-term use of the cave.
I feel that the drawings were probably the work of no more than a few
individuals, perhaps of a single individual over a period of time. Testing
this hypothesis will require painstaking study of microstylistic features
of the representations, but it is not an impossible task (See Muller 1977;
Hill and Gunn 1977). Unfortunately, we have little information about

decoration of house walls in the Southeast. Evidence for such wall decoration has recently been found at Toqua, a large Dallas culture village in the Little Tennessee River Valley in Tennessee, where several daub fragments were decorated with rows of conical punctuations and a painted fragment of wall daub was decorated with white dots and lines on a red wash (Polhemus 1985:21–22). Clay wall plaster generally could have been treated in a similar manner as the cave walls, if not, perhaps, with the same themes. As a result, the style of the Mud Glyph Cave stands out much more than it might have done if we had a view of the full range of eastern Tennessee prehistoric art. Until we have some better idea of the idiosyncrasy or social character of the Mud Glyph Cave art, it is difficult even to speculate whether this art represented private efforts to enhance spiritual power or some kind of ritual to protect the society as a whole.

Study of the execution of the drawings needs to be carried on to a finer level than is possible from the photographs and from the short visits that I and others have made to the cave. Because of variation in line depth and angle of lighting, photographs cannot always reveal sufficient detail to determine similarity of execution of sequence of action. On the other hand, the cave is so fragile that visitation has to be kept to the minimum consistent with adequate monitoring. One thing that can be done fairly well from the photographs is to investigate the way that the units are combined to form larger representations. Future studies along these lines are planned, and should help considerably in identifying similarities between the forms drawn in the cave and the symbolism known both prehistorically and historically from the Southeast.

The similarity of the themes of Mud Glyph Cave to those used in other styles in shell and copper has been noted; but, as also indicated, the style itself is quite distinct. In the discussion that follows, I make some comments on specific thematic similarities, but these do not imply stylistic similarity and do not necessarily indicate culture-historical connections between the artist(s) of the Mud Glyph Cave art and those of other areas. The themes, after all, were widely shared in Late Prehistoric times and may have been only generally similar in meaning across the East. The most common theme in the cave is that of the human figure, usually shown with what are likely to be wings hanging from outstretched arms. This is a pan-Eastern theme, but the Mud Glyph Cave versions are most like those from the engraved shell gorgets from its own region at sites like Big Toco (Thruston 1897:Plate XVIII). A number of differences can be noted, however. The cave art shows, for example, no clear examples of either antlers on the head or knotted animal skin attachments to the head that are seen in the shell gorgets (Phillips and Brown 1975–82:154). Similarly, the execution of claws and details of other parts of the body are very different. The serpent theme is rare in the thirteenth century in other surviving art from the eastern Tennessee Valley, although it becomes dominant by the fifteenth century and later. There is little general or specific similarity between the serpents of the Mud Glyph Cave and the later, so-called rattlesnake gorgets of the region. Both chevrons and concentric circle body units are critical design elements (at different times) in the rattlesnake gorgets but are lacking entirely in the Mud Glyph Cave. The greatest similarity of the Mud Glyph Cave serpents is to the roughly contemporary engraved serpent decorations on ceramics in central Alabama. This may suggest that the apparent absence of the serpent theme from the eastern Tennessee Valley in the thirteenth century is an artifact of selection of themes for representation in durable materials. As the cave shows, eastern Tennessee peoples were concerned with the "serpent" in earlier times, and the later dominance of the serpent theme in this region becomes more understandable, regardless of continuities or discontinuities in population and social identity.

Mud Glyph Cave is virtually unique in its being a body of prehistoric art from a relatively brief time span and in a medium that normally is not preserved in the archaeological record. Mud Glyph Cave may give a glimpse into beliefs of its time, but only as a result of very careful internal and comparative study.

B. BART HENSON **VII**

Art in Mud and Stone:
Mud Glyphs and Petroglyphs
in the Southeast

INTRODUCTION

Early travelers in the Southeastern United States noted in their journals the occasional rock art or glyphs they observed, and speculated upon their meaning. Native Americans consulted could offer little factual information as to the artists or their purpose in making the glyphs, but when they did venture an opinion it was usually to the effect that the glyphs were landmarks or related to spiritual matters.

Only the most exotic rock art sites came to the attention of early writers and historians; thus in the early literature one may find reference to only a small scattering of rock art sites in the Southeast. Garrick Mallery's (1893) voluminous survey of picture writing describes a few occurrences of rock art in the Southeast, two of which were portable glyph rocks. Steward (1937) in eloquent prose describes the many fantastic theories concerning the origin and explanations for petroglyphs, and science's efforts to hold in check "excesses of misinformation." As recently as 1946 it was reported that the states of Alabama, Mississippi, Florida, and South Carolina had no petroglyphs (Tatum 1946), and in 1967 it was reported that "fewer than 200 rock art sites have been found East of the Mississippi" (Grant 1967:15).

Recent literature and the numerous reports on file with institutions throughout the Southeast indicate that this number has been substantially increased, reflecting a vigorous and dynamic interest in the subject. It is only in the last decade that we have truly become aware of the quantities and varieties of rock art in mountaintop shelters, along valley trails, and in subterranean chambers of the southeastern United States.

In addition to the established archaeological societies and institutions engaged in rock art studies as one facet of their overall programs, associations and societies have been established and dedicated to the sole purpose of rock art study. The American Rock Art Research Association is representative of these organizations, is continental in scope, and publishes newsletters and journals related to its special interests. These organizations, by focusing attention on the subject, are bringing to light new information for scientific scrutiny.

In this paper, data from thirty-four rock art sites (Figure 4) studied by the author and in regional proximity with Mud Glyph Cave are examined to determine if a shared cultural affinity for particular motifs exists, and to determine if there exists a commonality of glyph characteristics. A large number of petroglyph sites have been discovered and described in Kentucky (see Coy and Fuller 1966, 1967, 1968, 1970, and 1971). Since the author has not, however, personally visited and studied these numerous sites, they are not included here. Particular emphasis is placed on Mud Glyph Cave and its possible relationship to other sites. Only sites with unmistakable glyphs, as opposed to utilitarian glyph-like features, have been used in trait comparisons.

The results of this review appear in Figure 5, while a brief narrative for each site provides a measure of explanation and background information. Plates XXI through XXIX represent a few glyphs from among many hundreds observed that appear to be predominately, and predictably, associated with the Mississippian cultural period. Some motifs are quite basic and are known to have been used for long periods of time, thus crossing cultural boundaries—stick figures, crosses, circles, and concentric circles being notable examples.

Although the author visited each site, many of them several times, and developed a photographic record of the art, it is apparent that a limit exists to observational accuracy, especially in glyphs with multitudinous superimpositions. Thus some traits, as tabulated in Figure 5, are subjective in nature.

The approximate geographic site locations, as shown in Figure 4, have

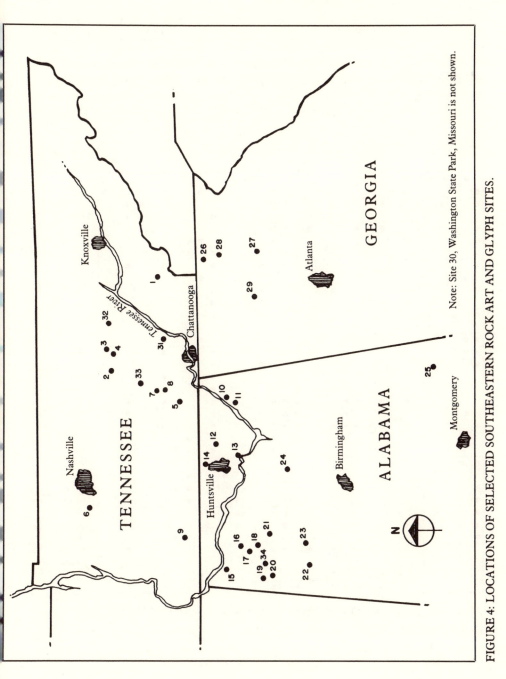

FIGURE 4: LOCATIONS OF SELECTED SOUTHEASTERN ROCK ART AND GLYPH SITES.

Note: Site 30, Washington State Park, Missouri is not shown.

MOTIFS							SITE RESIDUALS						TECHNIQUE						MEDIA					LOCATION								SITES		
aesthetic value	humans	animals	anthropomorphs	zoomorphs	shrine present	S.E.C.C. elements	quarrying	flint debris	pottery sherds	artifactual remains	mortars present	habitation obvious	bas-relief	drilled	pecked	rubbed	incised	painted	clay-mud	other stone	steatite	sandstone	limestone	open site	bedrock	boulders	cliff walls	shelter ceilings	shelter walls	cave ceilings	cave walls	COUNTY	STATE	NO.
	●	●	●	●	●	●											●		●												●	Monroe	TN.	1
●						●	●		●								●					●									●	White	TN.	2
●	●		●		●			●				●	●		●		●					●							●	●		White	TN.	3
														●		●	●					●							●	●		Van Buren	TN.	4
●			●		●			●	●	●		●		●		●	●					●							●	●		Franklin	TN.	5
●						●											●					●	●									Cheatham	TN.	6
●						●									●							●		●		●						Grundy	TN.	7
●						●							●	●	●			●				●							●		●	Grundy	TN.	8
●	●																●					●				●						Lawrence	TN.	9
●						●				●		●						●				●						●				DeKalb	AL.	10
						●		●										●				●					●					Jackson	AL.	11
●	●	●		●		●												●				●								●		Jackson	AL.	12
				●	●	●												●					●				●					Marshall	AL.	13
●	●									●		●	●									●			●	●						Madison	AL.	14
●	●	●				●				●		●			●	●						●				●						Colbert	AL.	15
●			●	●	●													●				●					●					Franklin	AL.	16
																	●					●					●					Lawrence	AL.	17
						●				●		●		●		●	●					●					●					Franklin	AL.	18
						●		●	●	●	●	●	●	●	●	·		●	●	●			●					●			●		Franklin	AL.	19
●	●	●			●	●						●		●	●		●	●	●			●								●		Marion	AL.	20
●		●				●		●	●	●	●	●	●	●	●		●	●				●					●			●		Lawrence	AL.	21
						●						●			●			●				●		●								Fayette	AL.	22
																	●					●		●	●							Walker	AL.	23
●															●							●		●	●							Blount	AL.	24
															●		●				●			●		●						Tallapoosa	AL.	25
●									●								●			●		●		●	●							Union	GA.	26
●																	●			●		●		●			●					Forsyth	GA.	27
●																	●			●		●		●			●					White	GA.	28
●						●											●					●		●		●						Cherokee	GA.	29
●						●											●					●		●	●							Washington	MO.	30
								●	●		●		●		●	●						●							●			Hamilton	TN.	31
						●											●						●							●		Cumberland	TN.	32
								●	●		●						●						●					●				Warren	TN.	33
								●	●	●	●	●		●			●					●		●			●					Marion	AL.	34

FIGURE 5: TABULATION OF TRAITS FOR 34 ROCK ART/GLYPH SITES IN THE SOUTHEAST.

been arbitrarily identified as Sites 1–34 here. Specific locations have been placed on file with the Department of Anthropology at the University of Tennessee, Knoxville.

SITE LOCATIONS AND USES

Rock art features usually appear in sheltered areas along or near trails and game routes, in places that favored the observation of valleys, and on open bedrock areas; a few are even found in caves. Many of these sites were probably utilitarian; i.e. used for rest stops, protection from the elements, and for hunting purposes. Other sites, not necessarily meeting these criteria, were nevertheless important to the Indian if the uniqueness or elaborateness of the glyphs are a valid indication. These latter sites have shamanistic implications and may have been used for performing hunting rituals or other socio-religious activities.

Caves in the Southeast have not been considered the domain of Indian artists; however, the discovery of Mud Glyph Cave (Site 1) in Tennessee in 1980 and its subsequent scientific study made a significant contribution to our knowledge and appreciation of the extent of aboriginal exploration and use of caves in the Southeast. When this discovery was announced, additional cave glyphs became known to researchers as residents from adjacent areas reported actual sites or gave information leading to the discovery of other such caves.

CAVES AND THEIR USES

For thousands of years Indians in the Southeast have explored caves and utilized them as sources of raw materials for tools and ornamentation, as mortuaries, and as sacred or ceremonial sites. The Indians also employed sheltered entrances of caves as habitation sites.

Indians in the Kentucky karst belt mined caves for gypsum, which was used to make ceremonial paint (Jackson 1982:34). Archaeological exploration of caves in the Mammoth Cave and Salts Cave systems in Kentucky has resulted in the discovery of a variety of artifactual materials suggesting both exploratory activities and mining for minerals and flint (Watson 1974:186). These activities were not isolated events, but were characteristic of caves in the region. The artifactual inventory resulting from these subterranean visits by Indians included cane fragments (torch material), charcoal (residue from torches and fires), gourd fragments, poles (for climbing), grass cord, dry reed stalks, wood, mining tools, and paleofecal fragments. Indian remains resulting from accidents and intentional interment have been found in Kentucky caves in both skeletal and mummy form (Robbins 1974:137).

Examination of footprints in Jaguar Cave, Tennessee, revealed that nine individuals, including two females and one child, traveled over 1,000 meters into the cave 4500 years ago (Robbins, Wilson, and Watson 1981:377). Footprints in a cave in Fentress County, Tennessee, have also been found, along with torch marks, charcoal, and evidence of high quality chert nodule mining (Ferguson 1983).

Cave burials have also been reported in Alabama, Tennessee, and Georgia (Walthall and DeJarnette 1974:1-2; Oakley 1971:85-9). Mortuary cave sites are predominantly of the Middle Woodland Period, utilized by people of the Copena culture, although skeletal remains in Crystal Onyx Cave in Barren County, Kentucky, have been dated to more than 2,500 years ago (Haskins 1983).

Ceremonial caves have been reported in other regions of the Americas. A remote cavern in Guatemala was recently discovered to contain tombs, pottery, stone work, and glyph paintings from the Mayan Classic Period of about 1,200 years ago (Witte and Garza 1981:149; Stuart 1981:220).

Eric S. Thompson in his introduction to a reprint edition of Henry C. Mercer's book, *The Hill-Caves of Yucatan*, quoting others, lists numerous cave burials, ossuaries, cremations, glyph chambers, mining activities, art galleries, and chambers for religious rites (1975). "The most sacred precincts of the Maya have always been caves. In these secluded grottoes the Maya's sacred fertility, calendrical and accession rituals were performed" (Pohl and Pohl 1983:28-9).

The dark mysterious recesses of caves in the Southeast must have attracted much curiosity from earliest times, just as they do today. In Mississippian times, caves possibly became places of religious and ritual activities. Underworld monsters were supposed to reside in caves, as caves were thought to be the entrances to the underworld in the later Historic Period. These beliefs created an awe and fear of caves, perhaps for a giant monster snake known to the Cherokees as Uktena. Snakes with wings, antlers, a horn, rattles, cat faces, and other embellishments were prominent motifs during the middle and late Mississippian Period.

Some investigators have described rock art sites as being private, semi-private, or public, depending upon their archaeological context (Docktor 1983:63). "Private" ritual sites are less likely to have bedrock mortars and more likely to be small and relatively inaccessible. "Public" ceremonial sites have essentially opposite characteristics—open, large, easily accessible, and possibly containing bedrock mortars.

TECHNIQUES

Rock art is created by altering a suitable surface by a single technique or combination of techniques. The terms "feature," "motif," "rock art," and "glyphs" are often used interchangeably to denote the subject. An explanation of some rock art terms:

"Rock Art"—a drawing or painting made on stone. In view of the glyphs of Mud Glyph Cave, the terms "Mud Glyph" may be added to the vocabulary to represent a drawing in mud.

"Pictograph"—a painting on stone. Pigments for the paint come from various natural sources of vegetable and mineral dyes. Hematite—yellow and red—was a popular pigment, as was charcoal.

"Petroglyph"—a drawing on stone, made by one of the following techniques:

Incising—cutting a line or groove in stone. Incising is usually found on soft surfaces where the art could be easily rendered, and was used to produce fine and careful detail.

Pecking—abrading a surface by striking with a hammer stone. Pecking is found on hard materials and was generally used to define a whole feature surface. Pecking may have been accomplished by both direct and indirect percussion techniques. Aesthetically executed glyphs may have been done by indirect percussion to attain the precise control necessary.

Rubbing—smoothing an incised or pecked feature by grinding or polishing with a harder stone.

Drilling—defining features by drilling a series of closely-spaced holes outlining the actual feature. This technique has been used both to define a feature, and to add emphasis to features created by another technique.

Bas-relief—a sculpture in low relief in which rock art projects only slightly from the surface, accomplished by removing the surrounding material.

DATING ROCK ART

Indian habitation sites in the Southeast are unequivocally radiocarbon dated as early as 6500–7600 B.C. (John W. Griffin 1974:14; DeJarnette 1962:1; Chapman 1976), and numerous artifacts exist to suggest their presence in the Southeast for millennia preceding this early period. Datable sites of great antiquity have usually been those cave and bluff shelters with well-defined stratigraphic sequences. Russell Cave, near the Tennessee River in northeast Alabama, and Stanfield-Worley Bluff Shelter in northwest Alabama are significant examples. These and other sites were occupied, intermittently it seems, from Early Archaic times through the late Mississippian Period. As man lived in these early sites in the Southeast, he probably created some form of art; i.e. glyphs, painting or sculpture indicative of his cultural attainments and beliefs. Although no forms of art dating to the Paleo or Early Archaic periods have been identified in the Southeast, this suggested artistic activity is entirely consistent with the much older sites throughout the world, notably in Europe (Leroi-Gourhan 1956), Africa (Lee and Woodhouse 1970), and Australia (Stubbs 1974). Thus art forms from the Paleo and Early Archaic periods could have existed on perishable media but apparently were not executed on the enduring stone surfaces in protected shelters and caves.

Attempts to date petroglyphs and pictographs by a variety of quasi-technical methods have proven to be generally unsatisfactory except for relative dating on a local level. Patination, lichen growth, rate of erosion, deposition covering glyphs, superimposition of glyphs, radiocarbon dating, association with datable artifacts, ethnographic accounts, and subject matter depicted are techniques cited as having potential for glyph date determination (Grant 1967:67; Schaafsma 1980:13).

Rates of patination, lichen growth, and erosion are highly variable and have not been determined for types of rocks or rock locations in the Southeast; thus attempts to use these methods are ill-advised. Direct radiocarbon dating of glyphs could only be applicable to those having a sufficient residue or organic binders from paint pigments—rare in the Southeast.

Ethnographic information relative to glyphs does not provide precise dates, other than the fact that the glyphs existed at the time of the ethnographic account. For example, the Track Rock Gap petroglyphs in north Georgia were said by the Cherokees to have been made by hunters for their own amusement while resting. Another tradition is that these were made while the surface of the newly created earth was still soft (Mooney 1900:418). These accounts are attributed to writers as early as 1834 and had a well-established oral tradition at that time. The actual age of the glyphs is unknown.

Pit-and-groove and concentric-circle petroglyphs associated with steatite quarrying operations observed in one open site can be dated, very tentatively by association, to the Late Archaic Period. A stump, left as steatite material was removed for bowl manufacture, is in one instance (Site 25) in direct association with a pit-and-groove glyph. As steatite vessel use seems to have been confined to the latter part of the Late Archaic Period, a date of ca. 2000 years B.C. is suggested. The same pit-and-groove type glyphs, however, are often in direct association with Mississippian glyphs.

The superimposition of one rock art motif over another can, logically, provide a clue to the relative ages of the figures. Glyphs from the majority of sites reported here seem to have a degree of homogeneity within the site, suggesting their origin as having derived from a single cultural group. If a site can be determined to have been used by only one culture, it may be assumed with some risk that any rock art found there was created by that culture. In those sites where several cultures have resided over long periods of time, dating rock art by assuming its association with datable artifacts is not possible alone; other data must be available.

While ceramics can be dated with reasonable accuracy, dating rock art by association with ceramics is risky unless there exists some other avenue of commonality. Other correlations can strengthen an assumed association: similar datable glyphs in the region, for example.

Comparisons of rock art motif style with those styles dominant in identifiable cultural periods have been the most satisfactory dating method in the Southeast. Also, certain stylistic elements or motifs in glyph/rock art forms can be identified as either present or absent at a particular site; these, along with other clues, can roughly date the rock art to a cultural period. Geographical location, too, can and does provide clues to a rock art site's probable cultural affiliation.

THE SOUTHEASTERN CEREMONIAL COMPLEX (SECC)

The Mississippian Culture, named for its beginnings in the Mississippi River Valley, flourished in the Southeast from about 700–1700 A.D. Until fairly recently the SECC, and indeed the Mississippian Culture, was thought by many to have been strongly influenced by Mexican and Middle American cultures (Howard 1968:3–8). Prevailing current opinion describes a Mississippian tradition, with its symbolism and artifactual inventory, having developed from prior resident cultural traditions (Phillips and Brown 1975:–82:20).

Archaeological remains of this culture are characterized by large truncated pyramidal mounds both individually and in major mound groups, especially in the Southeast, and elaborate shell-tempered ceramics. Major centers for the Mississippian Culture included areas now known as Moundville, Alabama; Etowah, Georgia; Spiro, Oklahoma; and Cahokia, Illinois, among many others. Manifestations of the culture's religious or ceremonial practices include distinctive motifs or symbols—cross, sun circle, bi-lobed arrow, mace, monolithic axe; animal forms—birds in variety, rattlesnakes in various forms; ceremonial objects—masks, baton or mace, monolithic axe, copper plates and badges, ceremonial flints; costume embellishments—antlered head-dress, skirt, feathers and bands on arms and legs; human forms—full figures, hands, feet, and heads.

This religion and its associated paraphernalia came to be known as the Southeastern Ceremonial Complex—SECC or the "Southern Cult." A paper published in 1945 provided the first detailed treatment of the SECC and its motifs (Waring and Holder 1945:9–29). In this work were categorized the various phenomena just described and, although now outdated, it is still considered a valuable source for SECC information.

SITE DISCUSSION

In the following paragraphs, brief site descriptions highlighting characteristics and locations supplement the tabular data of Figure 5. Sites 11-24 have appeared in the literature (Henson and Martz 1979:1-37) and are not specifically cited herein. Two items appearing in Figure 5 should be clarified—"shrine" and "aesthetics." The word shrine as used here represents a receptacle for sacred or religious symbols, and in a larger sense, a place having such symbols devoted to activities of a religious or ceremonial nature. Aesthetic determinations in rock art are essentially subjective evaluations by the viewer. Aesthetic values can be, and usually are, independent of cultural sophistication. For purposes of this summary, if the rock art/glyph is dimensionally symmetrical, appears to have been carefully planned, executed with care and precision, and perhaps enhanced by painting, or repeated with universal dimensional considerations, aesthetics are deemed to have been considered by the artist in and during their creation. If, on the other hand features are squiggles, random lines, scrawls, spaghettis/macaronis, or even reasonably-designed figures with randomness evidencing little or no planning or attention to execution, then aesthetics were probably not in the artist's mind.

SITE 1. MUD GLYPH CAVE

Mud Glyph Cave has been the subject of detailed study by an interdisciplinary team of experts whose work is presented elsewhere in this volume. It is presently the only cave known in the Southeast to have a major aboriginal ceremonial significance and to have received such an extensive scientific investigation of this activity.

Mud Glyph Cave, located in southeast Tennessee, is a relatively small cave, with a length of about 500 meters (see Chapter II). Boldly applied to mud deposited during ancient flooding, glyphs adorn the walls and ledges in the upper corridor of the mid portion of the cave in complex and bewildering quantities. These glyphs found in the permanently plastic mud, provided the name for this site, Mud Glyph Cave. Beyond the last mud glyph, a few incised petroglyphs appear on the limestone walls in a lower stream passage (Plate XXI).

The glyph gallery may be reached now with difficulty by wading and

PLATE XXI. Site 1, Petroglyphs, Mud Glyph Cave.

crawling through low and narrow passages; it was accessible to the Indians by the same general route as traveled today. It is a private place.

Glyphs were made by the simple expedient of applying one, two, or more fingers to the soft pliable clay mud and creating any design desired. Numerous impressions of fingers, and occasionally feet, are also present. In addition to the finger-made glyphs, sticks—probably cane—held individually and in clusters were widely used in creating the glyphs, some extending to depths of 5 mm. Snakes and serpentine lines, which may also represent snakes, are quite prevalent, while turtles whose bodies are represented by a heel print or a circular excision of mud are less frequent. Numerous zigzag lines and meanders are present as single lines or in pairs, triplets, and quadruples.

Mississippian Period SECC motifs are represented by the cross, circles, bi-lobed arrow, masks, human figure in costume dress, and in animal forms that include the rattlesnake, woodpecker, owl and hawk or eagle. Some of the classic SECC motifs, however, are absent at Mud Glyph Cave. The striped pole and apron are conspicuous by their absence here (see Chapter VI).

Fertility symbols, believed to be represented in the form of bisected circles or ovals, are prominent at Sites 3, 25, 29, and 30; however, no evidence of such symbols exist here. One small female stick figure with emphasized genitalia is present, but this does not seem to qualify as a fertility symbol. There is little, if anything, to suggest that these cave visitors were concerned with the concept of fertility rituals.

Mud Glyph Cave does not in general have art features aesthetically conceived or rendered. Rather it appears that the placement and execution of the glyphs were of much greater importance than aesthetic considerations. Actually, many glyphs were deliberately damaged, perhaps ritually, by smearing or striking with a pole or club. Ritual damage of cave art was noted in the famous Paleolithic caves of Europe, where both interior route and destination appeared to be part of the overall scheme of art gallery selection and development. In these instances, placement of the painting was apparently more important than the actual painting or its quality (Pfeiffer 1982:105–76), a trait possibly shared by the art in Mud Glyph Cave.

As this site is decidedly not utilitarian, not fertility oriented, and not aesthetically based but is possessed of innumerable glyphs; it seems logical to presume a special religious or ceremonial significance for the activities which took place there. The motifs present and a mean radiocarbon date of A.D. 1248 indicate that ceremonial activity peaked here in the thirteenth century A.D. Rock Art Sites 6, 8, 12, 18, 19, 20, 21, 30, and 32 are also believed to postdate A.D. 1200.

SITE 2, WHITE COUNTY, TENNESSEE

This is a cave site with many distinctive incised petroglyphs on its limestone walls; extensive superimposition of glyphs exists, and thus there are literally hundreds of glyphs present (Plate XXII). This cave is listed in Thomas C. Barr, Jr., *Caves of Tennessee* (1961:508), and the existence of the glyphs is noted. Projectile points and ceramic sherds from an occupation midden inside the cave suggest Mississippian Period association for the glyphs, although many glyphs are not particularly typical of this cultural period. The recent discovery of an incised horned serpent on the wall of this cave further strengthens the probability these glyphs date from the Mississippian Period. The cave has been investigated by the Department of Anthropology, University of Tennesseee at Knoxville, with a National Geographic Society grant (personal communication with Charles H. Faulkner).

SITE 3, WHITE COUNTY, TENNESSEE

Site 3 is located in a large, sandstone shelter at an elevation of approximately 1680 feet AMSL, having a south-southwest view. The soft sandstone walls of this shelter contain an extensive quantity of incised and drilled glyphs of abstract form. Vulva-like motifs are present in variety and quantity. One unusual feature is a creature incised on the core of a cane drilled hole, seemingly emerging from the hole. The hole periphery has lines radiating outward, possibly representing the mythical rebirth of the sun as it rises. Two masks appear on the ceiling, but these may be of historic origin.

SITE 4, VAN BUREN COUNTY, TENNESSEE

This is a sandstone bluff shelter site containing approximately 50 predominantly incised linear and rectilinear glyphs, and two vulva-like motifs. Many drilled holes exist, although no identifiable patterns are formed by the holes. The view is west-southwest at an elevation of approximately 1700 feet AMSL.

Mountain peaks of the same height are immediately in front of the shelter, thus its location provided no vantage point, and habitation is presumed to have been its principal use.

SITE 5, FRANKLIN COUNTY, TENNESSEE

This is a small sandstone shelter with predominantly pit and linear groove glyphs on walls and ceiling, with a view to the west. Its outstanding feature is a prepared elliptically shaped concavity, with peripheral rays, containing mythical anthropomorphic creatures with long insect-type antennae. Other significant features include patterns formed by drilled holes, and long linear features. This site appears to be a shrine.

Site 5 was excavated in the fall of 1974 by students and volunteers from the University of Tennessee at Chattanooga under the direction

PLATE XXII. Site 2, White County, Tennessee

of Duane King. The most significant discovery was chipped stone tools in an upper stratum of the site that had been heavily worn down from deeply incising the glyphs in the sandstone walls. A radiocarbon date from this stratum is A.D. 900 which indicates a Late Woodland/Early Mississippian association for these petroglyphs. A report on this site has never been published (personal communication with Charles H. Faulkner).

SITE 6, CHEATHAM COUNTY, TENNESSEE

The petroglyphs at this site are an incised mace and two closely associated appendages located on a high bluff overlooking Mound Bottom across a bend in the Harpeth River (Peacock 1949). The mace is oriented within 2° of the east-west direction and is approximately 27 inches in length (Plate XXIII). A Mississippian motif, it is believed to have been created by the same Mississippian people who built the truncated pyramidal mounds at the Mound Bottom site in the valley below.

SITE 7, GRUNDY COUNTY, TENNESSEE

A single maze of rectilinear design located at ground level on the flat surface of a bedrock boulder represents the site (Plate XXIV). The boulder is inclined slightly from the horizontal, and its location provides a view of a valley. Several small rock cairns, presumably containing burials, are located nearby.

SITE 8, GRUNDY COUNTY, TENNESSEE

Petroglyphs at this site are predominantly pecked crosses of Mississippian Period style in a high, but shallow, sandstone shelter, with considerable fall rock (Plate XXV). Other motifs include concentric circles and a birdman with spread wings. Elevation is approximately 1940 feet AMSL and the view is to the south-southeast. In addition to the more than 15 crosses, spirals, concentric circles, and a birdman on boulders in the shelter, three crosses appear on the rear wall. A vertical bluff wall approximately 75 feet above this shelter contains indistinct pictographic figures in red pigmentation.

SITE 9, LAWRENCE COUNTY, TENNESSEE

At this site the incised outline of a life-sized left hand and a figure resembling rattlesnake rattles or a medicine pipe appear in bedrock,

PLATE XXIII. Site 6, Cheatham County,
Tennessee.

adjacent to a small stream. No archaeological materials suggesting a possible cultural association were observed. The possibility of a historic origin
for these features has not been ruled out.

SITE 10, DEKALB COUNTY, ALABAMA

Pictographs at this site are located on the ceiling of a small sandstone
shelter with a southeast view. Five Mississippian Period motif pictographs
in yellow and red pigment are located on the ceiling—three crosses within
circles, a hand, and a crescent. This site has been archaeologically investigated and determined to have had an intermittent occupation for
approximately 9,000 years (Clayton 1967:1–35).

PLATE XXIV. Site 7, Grundy County, Tennessee.

SITE 11, JACKSON COUNTY, ALABAMA

Red pictographs are located on the vertical wall of a high bluff with a view to the east. Motifs include the spiral, horned snake, and rectangle. The site is a wide but shallow shelter, with a somewhat secluded location at an elevation which provides an overview of a valley and a tributary of the Tennessee River.

SITE 12, JACKSON COUNTY, ALABAMA

This site's motifs are pictographs located on the vertical face of the outside wall of a small sandstone shelter with a view to the east. Motifs are in red and yellow, five to seven feet above the shelter floor and include the spiral, eagle, snake, plumed snake, hand, mace, scalloped rectangle, and unidentified figures (Plate XXVI). The eagle now has been surreptitiously removed. Glyphs are Mississippian Period SECC (Cambron and Waters 1959:40).

SITE 13, MARSHALL COUNTY, ALABAMA

Pictographs at this site are located on the vertical face of a limestone bluff bordering the Tennessee River. This location provides a commanding view of the Tennessee River, both upstream and downstream. Motifs

PLATE XXV. Site 8, Grundy County, Tennessee.

PLATE XXVI. Site 12, Jackson County, Alabama.

painted in red include the circle, diamond, sun circle, bird, and other animal forms. The immediate area on both sides of the river was heavily occupied as early as the Archaic Period.

SITE 14, MADISON COUNTY, ALABAMA

A three-dimensional life-sized face is carved on the projecting point of a limestone rock at ground level. A circular mortar and a gourd-shaped mortar are located nearby. Early residents of the area have said at least three other faces were in existence as late as the turn of the century.

SITE 15, COLBERT COUNTY, ALABAMA

Glyphs representing human hands, feet, snakes, and other features are located on a boulder in a shelter floor. Rock art similar to this has been reported in Kentucky (Webb and Funkhouser 1932:224). Bedrock mortars are present, and two having a depth of more than 22 inches pass through a ground level boulder. Glyphs represent the Mississippian Period.

SITE 16, FRANKLIN COUNTY, ALABAMA

The glyphs at this site are located on the ceiling of a bluff shelter with a view to the east. Figures are painted in black and consist of anthropomorphs, zoomorphs, and animals which include the turtle, fox, and squirrel. This site appears on a game trail but is probably not related to hunting or hunting magic as such principal game species as the white-tailed deer are not represented.

SITE 17, LAWRENCE COUNTY, ALABAMA

The incised petroglyphs at this site are located on the flat surfaces of a sandstone boulder in a low shelter. About a dozen linear figures, possibly representing stick figures, are depicted.

SITE 18, FRANKLIN COUNTY, ALABAMA

Incised, pecked, and rubbed motifs at this site include concentric circles, circles with cup holes, spirals, cross, and parallel lines. A woodpecker and hand are also depicted, along with numerous meanders (Plate XXVII). These are located on sandstone boulders under a sandstone arch. Many of the petroglyphs are representative of the Mississippian Period (SECC) and in some instances are superimposed over other glyphs.

PLATE XXVII. Site 18, Franklin County, Alabama.

SITE 19, FRANKLIN COUNTY, ALABAMA
This site is a low, long sandstone bluff shelter facing east. Pecked and incised figures are located on a rear wall and on a large sandstone boulder in the shelter. Motifs include the cross, concentric circles, cup holes, parallel lines, and various meanders. Archaeological investigations have revealed this to be a multi-component site, with the Mississippian Period being heavily represented.

SITE 20, MARION COUNTY, ALABAMA
Rock art at this site represents the SECC at the height of its motif symbolism. Motifs include the mace, monolithic axe, several bi-lobed arrows, crescents, birds, animals, and human legs (Plate XXVIII). They were created by pecking and rubbing, and many still have traces of red pigmentation. The glyphs are on the walls of a small sandstone shelter facing east-southeast. The bi-lobed arrows, in contrast to many examples appearing on shell and copper, are pointed upward. Even two mono-lobed arrows are found at this site (Henson 1976:174–185). This site is secluded but accessible; bedrock mortars are present, and it was perhaps a public ceremonial site or shrine.

SITE 21, LAWRENCE COUNTY, ALABAMA
Petroglyphs at this site are located on a sandstone boulder near the front of a huge shelter. Motifs, predominantly Mississippian, are pecked and incised and include a cross in bas-relief, snake, sun with rays, animal figures, and meanders.

SITE 22, FAYETTE COUNTY, ALABAMA
Pecked glyphs at this site are located on a horizontal outcrop of sandstone bedrock. Motifs include a rattlesnake and a dog-like animal, plus mortar holes.

SITE 23, WALKER COUNTY, ALABAMA
Incised petroglyphs at this site are located on the flat roof of a sandstone shelter. Motifs include bow and arrow-shaped designs and various drilled and pecked hole patterns. No glyphs were observed on the shelter interior.

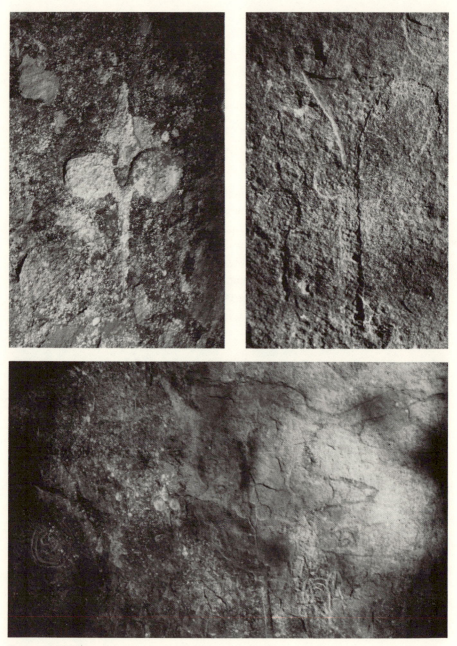

PLATE XXVIII. Site 20, Marion County, Alabama.

SITE 24, BLOUNT COUNTY, ALABAMA

The incised, pecked, and rubbed glyphs at this site are located on an exposed sandstone ridge overlooking a spring. Twenty-three cup-and-ring concentric circles, some clustered in groups, appear within a linear distance of 60 feet. Projectile points observed in the area are of the Woodland and Mississippian periods.

SITE 25, TALLAPOOSA COUNTY, ALABAMA

Glyphs at this site are pecked and incised in a steatite boulder in an open area. Motifs include pit-and-groove circles in association with stumps remaining from steatite bowl quarrying operations. A Late Archaic/Woodland association is presumed because of the intense activity here during this period.

SITE 26, UNION COUNTY, GEORGIA

This site is known as Track Rock Gap and consists of several large bedrock boulders with various deeply incised motifs. The principal feature appears to be the bisected ovals; but the rayed sun, cross within the circle, pit-and-groove and other features are present (Perryman 1962:19).

SITE 27, FORSYTH COUNTY, GEORGIA

This deeply incised boulder of granite was originally located in Forsyth County but now is on display at the University of Georgia campus in Athens. Concentric circles, pit-and-groove circles, "concentric circles with an acute angle," and pits are present (Jones 1873:377; Perryman 1961:1).

SITE 28, WHITE COUNTY, GEORGIA

Petroglyphs at this site appear on a granite boulder and include the encircled cross, ovals, circles, and tree-like figures. Originally from White County, this boulder is now located on the University of Georgia campus in Athens (Perryman 1962:21).

SITE 29, CHEROKEE COUNTY, GEORGIA

The petroglyphs are found on a granite boulder approximately 1.5 x 5 x 11 feet. Most motifs are concentric circles, ovals, encircled cross, pits, and unidentified features. The boulder has been removed from Cherokee County to the campus of Reinhardt College at Waleska, Georgia (Perryman 1962:75).

Site 30, Washington County, Missouri

This site is located in Washington State Park near DeSoto, Missouri. Petroglyphs numbering in the hundreds appear on outcrops of dolomite in an area of less than one acre. Glyphs include human figures, spirals, birds, footprints, mace, bi-lobed arrows, meanders, snakes, zoomorphs, phallic figures, and many unidentified figures (Diesing and Magre 1942:9). This is a large SECC glyph site, approximately 60 miles southwest of Cahokia.

Site 31, Hamilton County, Tennessee

Site 31 is a small sandstone shelter with a view to the south. Glyphs are concentrated on the left side of a shelter, three to six feet above the ground level. Motifs are predominantly linear and drilled. Lithic and ceramic debris present in the drip-line suggests a Late Woodland Period association.

Site 32, Cumberland County, Tennessee

At site 32, classic motifs of the SECC appear on the ceiling of a cave located at the head of the Sequatchie Valley. Incised glyphs include the encircled cross, mace, monolithic axe, woodpecker, crescent, and bird-man warrior (Plate XXIX). An incised circular figure, resembling a shield, is also present. Numerous charcoal smudge or stoke marks appear throughout the glyph area.

Site 33, Warren County, Tennessee

Glyphs at this site are located on two limestone boulders in a small sandstone shelter near a multicomponent, though predominantly Mississippian, site. All glyphs are linear, deeply incised, and range from a few inches to several feet in length. The site is too low and small to have been occupied or even used for shelter.

Site 34, Marion County, Alabama

Rock art at this site is located on the flat surfaces of large sandstone boulders in an open area bordering a creek. Rectangular incised figures, a human figure in headdress and mortar holes are present. The immediate

PLATE XXIX. Site, 32, Cumberland County, Tennessee.

area has produced lithic material from the Late Archaic Period through the Mississippian Period. Some of the features appear to be of historic origin.

CONCLUSIONS

The intricate detail and variety of the SECC motifs depicted in copper (Hamilton and Hamilton 1974) and in shell (Phillips and Brown 1975–82) are not found in petroglyph form because of the difficulty of implementing those designs in stone with the relatively crude tools available for stone working. Thus, only a small percentage of the actual SECC motif inventory appears as glyphs, and these are limited to the larger and simpler design elements. The limited number of the SECC motifs in rock and mud and their widespread use suggest that they may have had fairly concise definitions (Howard 1968:19). Many symbols are believed to relate to war or conjuring rather than to agriculture or fertility (Hudson 1976:88.

Mud Glyph Cave may be seen to correlate with sixteen of the sites surveyed in this study including the two caves, on the basis of motifs and temporal placement, but has a greater aura of ceremonialism than any other site. Its quantity and variety of glyphs are more extensive but less specialized than other classic SECC motif sites.

Traits from these thirty-four sites suggest the apparent preferences of the early artists for site location, glyph media, creative technique, site uses, motif elements, and aesthetic considerations. Most glyph sites are bluff shelters at the higher elevations and have incised and smoothed SECC motifs aesthetically rendered on boulders.

It is concluded that the extraordinary examples of rock art, and mud art, in the Southeast are Mississippian in orgin and represent a religious/ceremonial symbolism.

VIII

Prehistoric Cavers of
the Eastern Woodlands

Prior to the discovery of Mud Glyph Cave, the evidence for aboriginal exploration and use of caves in the eastern United States indicated a highly practical, even exploitative attitude toward subterranean natural resources. This pattern contrasts strongly with the evidence from Mud Glyph Cave for the nature of Mississippian activity there. My objective in this concluding chapter is to provide a context for evidence from Mud Glyph Cave by briefly summarizing, contrasting, and comparing prehistoric cave use elsewhere in the Midsouth, both earlier and contemporaneous.

The most abundant material for any time period comes from various parts of the Mammoth Cave system, but many other caves in the Midwest and Midsouth were also explored and used (see Table 3). In fact, rapidly accumulating information supports the generalization that every fairly dry cave with a reasonably accessible opening was entered and explored prehistorically, and that many such caves with special resources were systematically exploited.

CAVES AS PREHISTORIC
MINES AND QUARRIES

THE MAMMOTH CAVE SYSTEM, KENTUCKY

Some of the people who lived in the Mammoth Cave area of western Kentucky occupied the conspicuous natural entrances of what is now

TABLE 3. Summary Listing of Some Caves in the Midwest and Midsouth Known to Contain Evidence of Prehistoric Human Activity*

	Occupation of Entry Area	Disposal of the Dead	Exploration	Mining Cave Minerals	Mining Chert	Ceremonial
Big Bone Cave, Tennessee			X			
Copena Burial Caves, Alabama and Georgia		X				
Crystal Onyx Cave and Pit of the Skulls, Kentucky		X				
Fisher Ridge Cave, Kentucky			X			
Jaguar Cave, Tennesseee			X			
Lee Cave, Kentucky			X			
Mammoth Cave, Kentucky	X	X	X	X	X	
Mud Glyph Cave, Tennessee						X
Saltpeter Cave, Tennessee			X		X	
Salts Cave, Kentucky	X	X	X	X	X	
Sequoyah Cave, Alabama			X			
Short Cave, Kentucky	X	X				
Sinking Creek Cave, Kentucky	X	X			X	
Williams Cave, Virginia						X
Wyandotte Cave, Indiana			X	X	X	

*This is by no means an exhaustive listing, but is sufficiently comprehensive to indicate the range of prehistoric activity in caves of the Eastern Woodlands.

known to be the world's largest cave, the Mammoth Cave system (including both Salts Cave and Mammoth Cave as well as several others). They explored and mined the big cave interiors between 4,000 and 2,000 years ago (Brucker and R. Watson 1976; P. Watson et al. 1969; P. Watson, ed. 1974). They frequented areas several kilometers from the cave entrances and successfully negotiated a wide variety of complex and difficult passageways. These prehistoric cavers mined gypsum crystals as well as chert, and probably also two other minerals present in some abundance in the drier passages: epsomite (magnesium sulfate or Epsom salts) and mirabilite (sodium sulfate or Glauber's salt). Both crystalline salts have a cathartic effect on the human digestive system and each has a distinctive taste: epsomite tastes bitter and mirabilite tastes salty. Given the clear evidence for well-developed trade networks in the Eastern Woodlands by 2000 B.C. (Winters 1968; Goad 1978), I think it likely that these substances were passed on to other regions as well as being used locally.

WYANDOTTE CAVE, INDIANA

The exploitation pattern followed at Wyandotte Cave in southern Indiana (Munson and Munson 1981) is similar in some ways to that evidenced in the Mammoth Cave system, but exhibits interesting refinements, the most striking of which is the mining of a massive aragonite column (now known as "The Pillar of the Constitution") in a room 1 km from the entrance. This aragonite was probably traded throughout the Midwest (Tankersly et al. 1983). The Lost River or Wyandotte chert, also mined from Wyandotte Cave prehistorically, was doubtless another trade item.

There is clear evidence from Wyandotte Cave that the aboriginal people exploring and working there used shagbark hickory strips for illumination in contrast to the cane and dried weed stalk torches whose remains litter Salts Cave and Mammoth Cave. The contrast in torch material is probably because cane (*Arundinaria*) does not grow well as far north as Indiana. Dates on the archaeological remains in Wyandotte range from the early first millennium B.C. to the first few hundred years A.D.

SALTPETER CAVE, TENNESSEE

Another cave with abundant evidence for prehistoric mining activity is Saltpeter Cave in north central Tennessee, where large quantities of chert were dug from the sediments in a big underground room over 1000 meters from the entrance (Ferguson 1983). Besides quarrying chert nodules, the aboriginal miners undertook a considerable amount of preliminary working of the chert in the cave, so that there is an abundance of workshop debris on the cave floor and on the more or less level surfaces of large breakdown boulders in the quarry room. Radiocarbon dates for Saltpeter Cave range from 2400 B.C. to 800 B.C.

BIG BONE CAVE, TENNESSEEE

Intermittently during the nineteenth century, fragments of the skeleton of a giant sloth were found in a large, dry cave in Tennesseee that came to be called "Big Bone Cave." Henry Mercer of the University of Pennsylvania excavated parts of the cave in 1879 in an attempt to establish the relationship between the sloth remains and the abundant materials left by prehistoric human activities in the same passages (Mercer 1897). He was unable to document contemporaneity of sloth and human, and the archaeological as well as the paleontological remains were then ignored for nearly 100 years. During the summer of 1982, however, National Speleological Society cavers who were mapping the cave were sufficiently impressed with the quantity and state of preservation of the archaeological materials to contact archaeologists with experience in Tennessee and Kentucky caves.

As a result of their concern, an investigation of Big Bone Cave archaeology has been conducted by Charles Faulkner and George Crothers (Crothers 1986). Robert Stuckenrath and the Smithsonian Radiation Biological Laboratory have once again provided C^{14} dates as follows: 1615 ± 60 years: A.D. 335; 1595 ± 55 years: A.D. 355 and 440 ± 55 years: A.D. 1510. In addition, several dates from Beta Analytic document aboriginal activity in the cave during the first millennium B.C. (the six Beta dates range from 1050 B.C. to 170 B.C.). Hence, it appears that Big Bone Cave was explored at least intermittently for some 2,500 years. So far there is no clear evidence for mining of gypsum or other cave resources, but the cave has been severely disturbed by saltpeter mining and other recent use. It may be that most of the traces of aboriginal mining have been destroyed.

LEE CAVE

Lee Cave, inside Mammoth Cave National Park (Freeman et al. 1973), is another large cave that was partially explored prehistorically by local groups who scattered many fragments of charred cane along the main trunk passage (Marshall Avenue). There are large quantities of sulfate minerals in Lee Cave, as in the Mammoth Cave system, but no clear evidence of aboriginal mining. Epsomite is particularly abundant and— unlike the gypsum wall crust in Salts and Mammoth caves—could have been scooped up without leaving any long lasting traces. The quantity of torch debris in Marshall Avenue, however, does not indicate intensive or long term use. There is only one radiocarbon determination for the archaeological material in Lee Cave: 2250 B.C. ± 65.

FOOTPRINT CAVES

There is another category of cave archaeology in the eastern United States that might be called "footprint caves" (Watson 1983). Jaguar Cave, Fisher Ridge Cave, and Sequoyah Cave fall into this class. Archaeological remains comprise torch fragments (charred and uncharred), torch smudges on passage walls and ceilings, and prints in mud of the bare or slippered feet of the aboriginal cavers. There are approximately 274 complete footprints (of nine different people) in Jaguar Cave, about eighteen in Fisher Ridge Cave, and less than half-a-dozen in Sequoyah Cave. Footprint caves indicate little more than prehistoric interest in and exploration of local caves, although the abundance and excellent preservation of these traces in Jaguar Cave mean that some interesting physical anthropological information can also be derived from them (Robbins, Wilson, and Watson 1981). Radiocarbon dates (Table 4) indicate time spans for such exploration in ranges from ca. 2500 B.C. to A.D. 1430.

MORTUARY PITS AND CAVES

Another category of archaeological materials in caves features human remains. Some caves seem to have been used primarily as disposal places for the dead. This is the case for at least two caves on Prewitts Knob near Cave City, Kentucky; Crystal Onyx Cave and Pit of the Skulls (Haskins 1983). The one available radiocarbon date (on bone from Crystal Onyx Cave) is 680 B.C. Thus it appears that the people placing their dead in the pits on Prewitts Knob were at least partially contemporary with the major prehistoric caving activity in the Mammoth Cave system. The evidence, as presently available, consists of fragmentary skeletal remains (cranial and post-cranial) representing people of both sexes and all ages. The bodies were placed in, or in some cases thrown into, pits that have small openings on the surfaces of the Knob. No grave goods accompany the bones, but there are scatters of chert flakes including occasional tools

TABLE 4. Radiocarbon Dates from Footprint Caves
in the Eastern Woodlands*

Fisher Ridge Cave, Kentucky	800 B.C. ± 85
	1225 B.C. ± 80
Jaguar Cave, Tennessee	2580 B.C. ± 85
	2680 B.C. ± 75
	2745 B.C. ± 85
Sequoyah Cave, Alabama	A.D. 1430 ± 50

*All dates were provided through the kind offices of Dr. Robert Stuckenrath and the Radiation Biology Laboratory of the Smithsonian Institution, Rockville, Md. Uncalibrated, Libby half-life, 1950 base date.

(projectile points and retouched flakes, for example), in various places on the Knob surface. The chert nodules that furnished the raw material for these items also outcrop on the Knob.

Several aboriginal bodies were found by saltpeter miners in Short Cave (which is adjacent to but outside the southern border of Mammoth Cave National Park) during the nineteenth century. Details are lacking for these finds (see Meloy and Watson 1969; Meloy 1971), but what information is available suggests a late prehistoric placement for at least one of the bodies, which was laid away in a stone-box or stone-slab grave. Similarly, in several Alabama and Georgia caves, the honored dead were ceremoniously interred some 1500 years ago during what is now known as the Copena Period (Walthall and DeJarnette 1974).

Much more recently (1982–1983), vandals looted the remains of several prehistoric graves in Sinking Creek Cave near Bowling Green, Kentucky (Wilson 1982; Hensley-Martin 1986). This site is somewhat like Short Cave in that it consists of a short piece of trunk passage, open at both ends, that was used as a chert quarry, and as a habitation and burial place by some members of the local prehistoric population. There are no available radiocarbon determinations for Sinking Creek Cave as yet, but the archaeological materials indicate an approximate time span of several thousand years B.C. down to about 1,500 years ago.

CEREMONIAL CAVES

MUD GLYPH CAVE, TENNESSEE

In contrast to what we know about the business-like aspect of aboriginal activities in many of the caves just briefly described, Mud Glyph Cave offers clear evidence for ritual or ceremonial use. Although Mud Glyph is a rather small cave, it requires some effort and special equipment (illumination at least) to enter and to explore. As documented in detail in this volume, most of the people who went into Mud Glyph Cave seven centuries ago apparently did so with only one purpose in mind: to communicate with, or to approach, the supernatural. The representations they created on the mud-coated cave walls were surely important to them, and the primary function of these tracings was surely ideological. Hence, use of Mud Glyph Cave is significantly different from what we know of cave use nearly everywhere else in the eastern United States.

So far there are only two other documented examples of prehistoric ceremonial caves in the Eastern Woodlands that are not also habitation sites, mortuary sites or quarries (Williams Cave in Virginia, with radiocarbon determinations clustering around A.D 1100, and the cave in Cumberland County, Tennessee, Site 32, Chapter VII, this volume), but there must have been at least a few others. At any rate, however, there does seem to be a contrast emerging between an earlier period of cave use (roughly, Late Archaic to early Middle Woodland; ca. 2000 B.C. to A.D. 300 or 400) and a later period (later Middle Woodland to Mississippian; ca. A.D. 300 or 400 to A.D. 1500). In the earlier period, caves in many places in the Eastern Woodlands were obviously used intensively as sources of desirable minerals and chert. In the later period they served as burial sites or—perhaps—as contact points with the underworld. If such a contrast is substantiated by further research in cave archaeology, then we must try to understand and explain why such a major shift occurred from use of (and presumably conception of) caves as important—but relatively straightforward or prosaic—economic resources to caves as having primarily or solely supernatural significance. An explanation for the shift will obviously have to include some detailed discussion of the role played by the underworld and its inhabitants (such as the Uktena [Hudson 1976:144–147]) in the mythology of the human groups who were living in this part of the New World when Europeans first arrived. As clearly shown elsewhere in this book, there are resemblances between some of the art motifs on the walls of Mud Glyph Cave and some of the basic themes in the mythology of the Five Civilized Tribes of the Southeastern United States. An explanation of Mud Glyph Cave, therefore, necessitates attention to the descendents of its creators as well as to their contemporaries (the Late Woodland and Mississippian peoples of eastern Tennessee) and their ancestors: the Archaic and Woodland cavers of Tennessee, Kentucky, and elsewhere in the Midwest and Mid-south of the United States.

REFERENCES CITED

Barbour, Roger W., and Wayne H. Davis
  1974   *Mammals of Kentucky.* Univ. Press of Kentucky, Lexington.
Barr, Jr., Thomas C.
  1961   *Caves of Tennessee.* State of Tennessee Dept. of Conservation and Commerce, Division of Geology, Bulletin 64, Nashville.
Bartram, William
  1928   [1791] *Travels Through North & South Carolina, Georgia, East & West Florida . . . .* Reprinted as *Travels of William Bartram,* ed. Mark Van Doren. Dover Press, New York.
Berner, Alfred, and Leslie W. Gysel
  1967   Raccoon Use of Large Tree Cavities and Ground Burrows. *Journal of Wildlife Management* 31(4):707-14.
Black, Glenn A.
  1967   *Angel Site: An Archaeological, Historical, and Ethnological Study.* Indiana Historical Society, Indianapolis.
Brucker, Roger W. and Richard A. Watson
  1976   *The Longest Cave.* Knopf, New York.
Caldwell, Joseph, and Catherine McCann
  1941   *Irene Mound Site, Chatham County, Georgia.* Univ. of Georgia, Athens.
Cambron, James W., and Spencer A. Waters
  1959   Petroglyphs and Pictographs in the Tennessee Valley and Surrounding Area. *Journal of Alabama Archaeology* 5(2):27-51.
Chapman, Jefferson
  1976   The Archaic Period in the Lower Little Tennessee River Valley: The Radiocarbon Dates. *Tennessee Anthropologist* 1(1):1-12.
Clayton, Margaret V.
  1967   Boydston Creek Bluff Shelter Excavations. *Journal of Alabama Archaeology* 13(1):1-35.
Coy, Fred E., Jr., and Thomas C. Fuller
  1966   Petroglyphs of North Central Kentucky. *Tennessee Archaeologist* 22(2):53-66.
  1967   Turkey Rock Petroglyphs, Green River, Kentucky. *Tennessee Archaeologist* 23(2):58-79.
  1968   Tar Springs Petroglyphs, Breckinridge County, Kentucky. *Tennessee Archaeologist* 24(1):29-35.
  1970   Reedyville Petroglyphs, Butler County, Kentucky. *Central States Archaeological Journal* 17(3):101-109.
  1971   Petroglyphs of Powell County, Kentucky. *Central States Archaeological Journal* 18(3):113-22.
Crane, H.R., and James B. Griffin
  1961   University of Michigan Radiocarbon Dates VI. *American Journal of Science Supplement* 8:105-125. New Haven.

Crothers, George
1986   *Final Report on the Survey and Assessment of the Prehistoric and Ar-
       cheological Remains in Big Bone Cave, Van Buren County, Tennessee.*
       Report submitted to the Tennessee Department of Conservation.
       Department of Anthropology, University of Tennessee, Knoxville.
DeJarnette, David L., Edward B. Kurjack, and James W. Cambron
1962   Stanfield-Worley Bluff Shelter Excavations. *Journal of Alabama Ar-
       chaeology* 8(1 and 2).
Diesing, Eugene H., and Frank Magre
1942   Petroglyphs and Pictographs in Missouri. *The Missouri Archaeologist*
       8(1):9–18.
Docktor, Desiree
1983   The Significance of Rock Art Setting in the Interpretation of Form
       and Function: Preliminary Investigation of Two Yokuts Rock Art Sites
       in California. In *American Indian Rock Art Volume 9*, ed. Frank G.
       Bock, pp. 63–71. American Rock Art Research Association, El Toro,
       Calif.
Doutt, J. Kenneth, Carolina A. Heppenstall, and John E. Guilday
1967   *Mammals of Pennsylvania.* Pennsylvania Game Commission, Har-
       risburg.
Faulkner, Charles H.
1983   Personal communication.
1986   A Study of Seven Southeastern Glyph Caves. Final report submitted
       to the National Geographic Society.
Faulkner, Charles H., Bill Deane, and Howard H. Earnest, Jr.
1984   A Mississippian Period Ritual Cave in Tennessee. *American Antiquity*
       49(2):350–61.
Ferguson, Lee G.
1983   An Archaeological Investigation of TCS #FE60: A Cave in North Cen-
       tral Tennessee. Proceedings of the National Speleological Society An-
       nual Meeting, June 27–July 1, 1983 (Abstract). *The NSS
       Bulletin* 45(2).
Freeman, J.P., G.L. Smith, T.L. Poulson, P.J. Watson, and W.B. White
1973   Lee Cave, Mammoth Cave National Park, Kentucky. *The NSS Bulletin*
       35:109–26.
Goad, Sharon I.
1978   Exchange Networks in the Prehistoric Southeastern U.S. Unpubl.
       Ph.D. diss., Univ. of Georgia, Athens.
Golley, Frank B.
1962   *Mammals of Georgia. A Study of Their Distribution and Functional Role
       in the Ecosystem.* Univ. of Georgia Press, Athens.
Grant, Campbell
1967   *Rock Art of the American Indian.* Thomas R. Crowell, New York.
Griffin, James B.
1963   A Radiocarbon Date on Prehistoric Beans from Williams Island,
       Hamilton County, Tennessee. *Tennessee Archaeologist* 19(2):43–46.
Griffin, John W.
1974   *Investigations in Russell Cave.* National Park Service, U.S. Dept. of
       the Interior, Washington, D.C.

Hall, E. Raymond, and Keith R. Kelson
1959   *The Mammals of North America*. Ronald Press, New York.
Hamilton, Henry W., and Jean Tyree Hamilton
1974   *Spiro Mound Copper*. Missouri Archaeological Society Memoir No. 11.
Haskins, Valerie A.
1983   The Archaeology of Prewitts Knob, Kentucky. Proceedings of the National Speleological Society's Annual Meeting June 27–July 1, 1983 (Abstract). *The NSS Bulletin* 45(2).
HensleyMartin, Christine
1986   The O'Bryan Cave Site, 15Si9: An Early Archaic to Early Woodland Occupation Site in the Central Kentucky Karst. Unpubl. Master's thesis, Washington Univ., St. Louis.
Henson, B. Bart
1976   A Southeastern Ceremonial Complex Petroglyph Site. *Journal of Alabama Archaeology* 22(2):175–85.
Henson B. Bart, and John Martz
1979   *Alabama's Aboriginal Rock Art*. Alabama Historical Commission, Montgomery.
Hill, James, and J. Gunn (editors)
1977   *The Individual in Prehistory: Studies in Variation in Style in Prehistoric Technologies*. Academic Press, New York.
Howard, James H.
1968   *The Southeastern Ceremonial Complex and its Interpretation*. Missouri Archaeological Society Memoir No. 6.
Hudson Charles
1976   *The Southeastern Indians*. Univ. of Tennessee Press, Knoxville.
Jackson, Donald Dale
1982   *Underground Worlds*. Time-Life Books, Alexandria, Va.
Jones, Jr., Charles C.
1873   *Antiquities of the Southern Indians* (1972 reprinting). Reprint Company, Spartanburg, S.C.
Kneberg, Madeline
1959   Engraved Shell Gorgets and their Associations. *Tennessee Archaeologist* 15(1):1–39.
Lee, D.N., and H.C. Woodhouse
1970   *Art on the Rocks of Southern Africa*. Scribner's, New York.
Leroi-Gourhan, Andre
1956   *Treasures of Prehistoric Art*. Harry J. Abrams, New York.
Linzey, Alicia V., and Donald W. Linzey
1971   *Mammals of Great Smoky Mountains National Park*. Univ. of Tennessee Press, Knoxville.
McGimsey, Charles R.
1964   An Engraved Shell Gorget from the McDuffee Site. *Bulletin of the Arkansas Archaeological Society* 5(7):128–29.
Mallery, Garrick
1893   Picture-Writing of the American Indian. *Tenth Annual Report of the Bureau of Ethnology*, 1888–89, Government Printing Office, Washington, D.C.

Meloy, H.
 1971   *Mummies of Mammoth Cave.* Micron, Shelbyville, Ind.
Meloy, H., and Patty J. Watson
 1969   Human Remains: "Little Alice" of Salts Cave and Other Mummies. In The Prehistory of Salts Cave, Kentucky, by P.J. Watson et al. Illinois State Museum, *Reports of Investigations* No. 16.
Mercer, Henry C.
 1897   The Finding of the Remains of the Fossil Sloth at Big Bone Cave, Tennessee, in 1896. *Proceedings of the American Philosophical Society* 36(154):36–70. Philadelphia.
 1975   *The Hill Caves of Yucatan*, with a new introduction by Sir J. Eric S. Thompson. Univ. of Oklahoma Press, Norman, pp. vii–xliv.
Mooney, James
 1900   *Myths of the Cherokee.* Bureau of American Ethnology, 19th Annual Report. Washington, D.C.
Moore, Clarence B.
 1915   Aboriginal Sites on Tennessee River. *Journal of the Academy of Sciences of Philadelphia* XVI:169–428.
Muller, Jon
 1966a Archaeological Analysis of Art Styles. *Tennessee Archaeologist* 22(1):25–39.
 1966b An Experimental Theory of Style. Unpub. Ph.D. diss., Harvard Univ., Cambridge.
 1970   Review of *The Southeastern Ceremonial Complex and Its Interpretation* by James H. Howard. *American Anthropologist* 72(1):182–83.
 1971   Style and Culture Contact. In *Man Across the Sea: Problems of Pre-Columbian Old World-New World Contacts*, ed. C. Riley, et al., pp. 66–78. Univ. of Texas Press, Austin.
 1973   Structural Studies of Art Styles. IXth International Congress of Anthropological and Ethnological Sciences, No. 1036.
 1977   Individual Variation in Art Styles. In *The Individual in Prehistory: Studies in Variability in Style in Prehistoric Technologies*, ed. J. Hill and J. Gunn, pp. 23–29. Academic Press, New York.
 1979   Structural Studies of Art Styles. In *The Visual Arts: Plastic and Graphic*, ed. J. Cordwell, pp. 139–211 (& pp. 10–20). Mouton, The Hague.
 1983   The Southeast. In *Ancient North Americans*, ed. J. Jennings, pp. 372–419. W.H. Freeman, San Francisco.
 1983   Review of *Stylistic Variation in Prehistoric Ceramics* by Stephen Plog. *American Antiquity* 48(1):195–96.
 1984   Review of *Prehistoric Shell Engravings from Spiro* by P. Phillips and J. Brown. *American Antiquity* 49(3):669–670.
 In   The Southern Cult. Volume 13, *Handbook of North American Indians*, press   *The Southeast*, ed. W. Sturtevant. Smithsonian Press, Washington, D.C.
Munson, Cheryl A., and Patrick J. Munson
 1981   Archaeological Investigations, 1980: Wyandotte Cave, Ind. Prepared for the Division of Forestry, Indiana Dept. of Natural Resources.
Munson, Patrick J., and James H. Keith
 1984   Prehistoric Raccoon Predation on Hibernating *Myotis*, Wyandotte Cave, Ind. *Journal of Mammalogy* 65(1):152–55.

Murie, Olaus J.
1975   *A Field Guide to Animal Tracks*. Houghton Mifflin, Boston.
Oakley, Jr., Carey B.
1971   *An Archaeological Investigation of Pinson Cave (Je20))*. Univ. of Alabama, Tuscaloosa.
Peacock, Charles K.
1949   Mound Bottom Pictograph. *Tennessee Archaeologist* 5(1):8–9.
Perryman, Margaret
1961   Sculptured Monoliths of Georgia. *Tennessee Archaeologist* 17(1):1–9.
1962   Sculptured Monoliths of Georgia–Part 2. *Tennessee Archaeologist* 18(1):15–22.
Pfeiffer, John E.
1982   *The Creative Explosion*. Harper and Row, New York.
Phillips, Philip, and J.A. Brown
1975–   *Pre-Columbian Shell Engravings from the Craig Mound at Spiro,*
82   *Oklahoma.* (6 vol.) Peabody Museum Press, Cambridge, Mass.
Pohl, Mary, and John Pohl
1983   Ancient Maya Cave Rituals. *Archaeology* 36(3):28–32; 50–51.
Polhemus, Richard R.
1985   Mississippian Architecture: Temporal, Technological, and Spatial Patterning of Structures at the Toqua Site (40MR6). Unpubl. Master's thesis, Univ. of Tennessee, Knoxville.
1986   The Toqua Site—40MR6, A Late Mississippian Dallas Phase Town. Final report for the National Park Service and Tennessee Valley Authority.
Robbins, Louise M.
1974   Prehistoric People of the Mammoth Cave Area. In *Archeology of the Mammoth Cave Area*, ed. Patty Jo Watson, pp. 137–62. Academic Press, New York.
Robbins, Louise M., Ronald C. Wilson, and Patty Jo Watson
1981   Paleontology and Archaeology of Jaguar Cave, Tennessee. In *Proceedings of the VIIIth International Congress of Speleology, Vol. I*, ed. B. Beck, pp. 377–80.
Rouse, Irving
1951   A Survey of Indian River Archaeology, Florida. *Yale University Publications in Anthropology* No. 44.
Sagan, Carl
1980   *Cosmos*. Random House, New York.
Salo, Lawr V. (editor)
1969   *Archaeological Investigations in the Tellico Reservoir, Tennessee, 1967–69: An Interim Report*. Univ. of Tennessee, Dept. of Anthropology, Knoxville.
Schaafsma, Polly
1980   *Indian Rock Art of the Southwest*. Univ. of New Mexico Press, Albuquerque.
Schroedl, Gerald F.
1978   Excavations of the Leuty and McDonald Site Mounds. Univ. of Tennessee, Dept. of Anthropology, *Report of Investigations* No. 22 and TVA *Publications in Anthropology* No. 15. Chattanooga.
Stewart, Julian H.
1937   Petroglyphs of the United States, *Annual Report* of the Board of Regents

of the Smithsonian Institution, Publication 3405, pp. 405–25. Government Printing Office, Washington, D.C.

Stuart, George E.
1981 Maya Art Treasures Discovered in Cave. *National Geographic* 160(2):220–35.

Stubbs, Dacre
1974 *Prehistoric Art of Australia*. Scribner's, New York.

Tankersley, Kenneth B., Patrick J. Munson, and Cheryl Ann Munson
1983 Physical and Chemical Properties of Wyandotte Cave Aragonite and Their Archaeological Implications: Preliminary Findings. Paper submitted to *Geo² The Journal of the NSS Section of Cave Geology and Geography*.

Tatum, R.M.
1946 The Importance of Petroglyphs in Tennessee. *Ten Years of the Tennessee Archaeologist: Selected Subjects*. J.B. Graham Publisher, Chattanooga, pp. 40–41.

Thruston, Gates P.
1897 *The Antiquities of Tennessee*. Cincinnati.

Walthall, John A., and David L. DeJarnette
1974 Copena Burial Caves. *Journal of Alabama Archaeology* 20(1):1–62.

Waring, Jr., A.J., and Preston Holder
1945 A Prehistoric Ceremonial Complex in the Southeastern United States. Reprinted from *American Anthropologist*, Vol. 47, No. 1 in *The Waring Papers*, ed. Stephen Williams, pp. 9–29. Peabody Museum, Cambridge, Mass.

Watson, Patty Jo
1983 Prehistoric Footprints in United States Caves. Paper presented at the National Speleological Society Convention, Elkins, West Va.
1969 The Prehistory of Salt's Cave, Kentucky. Illinois State Museum, *Reports of Investigations* No. 16.
1974 Observations in Upper and Lower Mammoth. In *Archeology of the Mammoth Cave Area*, ed. Patty Jo Watson, pp. 185–92. Academic Press, New York.

Watson, Patty Jo (editor)
1974 *Archeology of the Mammoth Cave Area*. Academic Press, New York.

Webb, William S., and W.D. Funkhouser
1932 *Archaeological Survey of Kentucky*. Dept. of Anthropology and Archaeology, Univ. of Kentucky, Lexington.

Wilson, Ronald C.
1982 Notes on a Visit to the Sinking Creek Cave System, Simpson County, Kentucky. December 4, 1982. Ms. in possession of author.

Winters, Howard D.
1968 Value Systems and Trade Cycles of the Late Archaic in the Midwest. In *New Perspectives in Archaeology*, ed. S.R. Binford and L.R. Binford, pp. 175–221. Aldine, Chicago.

Witte, Karen T.A., and E. Garza
1981 NAJ TUNICH—The Writing on the Wall. *National Speleological Society News*, pp. 149–51, National Speleological Society.

INDEX

The Prehistoric Native American Art of Mud Glyph Cave was composed into type on the Compugraphics MCS 100 digital phototypesetter in ten point Plantin with three points of spacing between the lines. Plantin was also used for display. The book was designed by Matt Williamson, typeset by BGA Graphics, printed offset by Thomson-Shore, Inc. and bound by John H. Dekker & Sons. The paper on which the book is printed is designed for an effective life of at least three hundred years.

THE UNIVERSITY OF TENNESSEE PRESS: KNOXVILLE